future face

image, identity, innovation

sandra kemp

with contributions from
vicki bruce and alf linney

P
PROFILE BOOKS

First published in Great Britain in 2004 by
PROFILE BOOKS LTD
58a Hatton Garden
London EC1N 8LX
www.profilebooks.co.uk

in association with The Wellcome Trust

10 9 8 7 6 5 4 3 2 1

A CIP catalogue record for this book is available
from the British Library.

ISBN 1 86197 768 9

Editorial consultant: Paul Forty
Text and cover design: Studio Myerscough
Printed and bound in Great Britain
by St Edmundsbury Press, Bury St Edmunds

contents

author's acknowledgements

Research for this book and an exhibition on the same theme has extended over the past five years. I have had immense pleasure in working with a number of people and in libraries, galleries, museums and archives in Britain, Europe, North America and Australia, and I would like to thank everyone who has contributed to *Future Face*.

I am particularly grateful to my research assistant, Georgia Dickson, for her hard work, energy and resourcefulness. One of my happiest memories of the project is time spent at Blythe House, the Science Museum and Wellcome object archive, armed with a handwritten catalogue and digital camera. At different stages of the project, Andrew Bamji, Eleanor Crook, Catherine Hemelryk, Cathy Johns, Rachel Kent, Anna Lobbenberg, Yve Lomax, Julienne Lorz, Alexandra MacGilpin, Katie Maras, Richard Neave, Alice Nicholls, Mary Panzer, Sarah Simblet and Helen Tysoe also provided specialist research assistance.

A number of people have provided invaluable help by reading the manuscript of the book for me. I am grateful to Mark Hanson, Roger Hargreaves, Sarah Kember, Giles Newton, Ian Jones, Tim Boon, David Trotter and Robert Sobieszek for their comments and suggestions.

The face has been the subject of many and wide-ranging studies. I have detailed the books and exhibitions that I have found most illuminating in an annotated select bibliography. Robert A. Sobieszek's *Ghost in the Shell* and Peter Hamilton and Roger Hargreaves's *The Beautiful and the Damned* are extraordinary investigations of the face in photographic portraiture (and each book also accompanied an exhibition). Deanna Petherbridge's groundbreaking exhibition on art and anatomy, *The Quick and the Dead*, which worked closely with material from both art and science, was inspirational for my approach to the material. David Sylvester's interviews with the artist Francis Bacon, together with the neuro-physiologist Jonathan Cole's interviews with people whose experience of faces had significantly changed their lives, offer a different – and sometimes shocking – route into the relationship between face and identity. Kobo Abe's *The Face of Another*, about a man who has lost his face, and with it his connection to other people, is one of the most powerful novels I have read.

The five main institutions involved with *Future Face* – the National Portrait Galleries in London and at the Smithsonian Institution, Washington, the Science Museum in London, the Wellcome Trust and the Royal College of Art – have provided enormous support. John Cooper and Marc Pachter have nourished my fascination with and knowledge of portraiture. I started work on *Future Face* at the Smithsonian, and Marc and I have continued talking (and arguing) about faces and portraits ever since. Andrew Nahum has been behind the project from start to finish. I am a great admirer of his own exhibition work. Christopher Frayling persuaded me to continue when other work commitments threatened to overwhelm me and has provided full Royal College of Art support. Ken Arnold has project-managed the exhibition and supported the book as well. Morag Myerscough has added a new dimension by thinking through the content from a design perspective. It has been a pleasure working with her. I am grateful to Margaret Blunden at the University of Westminster and Michael Worton at University College London for their support over a number of years. I shall miss Margaret when she retires in 2005. She has been an inspirational colleague and friend.

Generous funding from the Smithsonian Institution, Washington, the Engineering and Physical Sciences Research Council (EPSRC), the British Psychological Society, and the Royal College of Art has made the research possible. Working with my collaborators, Vicki Bruce and Alf Linney, has been one of the high points of *Future Face* and led to a year of close interdisciplinary networking and research. Each of them is a pioneer in their own field.

In my own office Virginia Ferreira and Vanessa Humphreys have created an atmosphere of friendship and support that allowed me to focus on *Future Face* when there was always really too much else to do. Without them, neither the book nor the exhibition would have happened.

I would also like to thank my editors, Jane Hogg at the Wellcome Trust and Paul Forty at Profile Books.

My most heartfelt thanks are to Lisa and Peter Lewis and to Jack Kemp for emergency scanning of images and encouragement, and for putting up with *Future Face* for far too long. This book is dedicated to them.

Sandra Kemp
September 2004

introduction

sandra kemp

Milan Kundera's novel *Immortality* contains a number of meditations on the meaning of the face. In the following extract, two characters, Agnes and Paul, are discussing the relationship between the face and identity, and Agnes says:

'Yes, you know me by my name, you know me as a face and you never knew me in any other way. Therefore it could never occur to you that my face is not myself. Yet, there comes a time when you stand in front of a mirror and ask yourself: this is myself? And why? Why did I want to identify with this? What do I care about this face? And at that moment everything starts to crumble.'

It was these kinds of questions that fascinated me in researching and assembling the material for this book. I wanted to find out more about the complex role that faces play in our lives and what they tell us about ourselves. I centred my investigation on three questions. What is a face? How do we portray and interpret faces? What will faces look like in the future?

Future Face is published to coincide with the opening of the exhibition on the same theme and with the same title, which I have curated and which can be seen at the Science Museum in London from October 2004 until February 2005. The CD-ROM catalogue of the exhibition is available on request and free of charge to readers of this book, providing an opportunity to supplement the book's text and images. (For details of how to apply for the CD-ROM, see page 224.) The scientific, interactive and object-rich focus of the CD-ROM is designed to complement the narrative material here, which in turn draws on and develops the themes of the exhibition.

the face

'Every age has its own gait, glance and gesture,' wrote Charles Baudelaire in *The Painter of Modern Life* (1863). The images of two faces are central to my project. The first, is perhaps *the* 21st-century face: that of Michael Jackson. Race is one of the many ways in which we attribute meaning to the face and the adoption of a new or different cultural facial style is a powerful indication that the face is not universal. Our notions of the face are subtly tinged or tainted by

Franz Xavier Messerschmidt
The Yawner (detail), circa 1777–83

Mug-shot of Michael Jackson, 2003 (p.10)
Wladyslaw Theodore Benda,
Lifemask of Myrna Loy, circa 1940 (p.11)

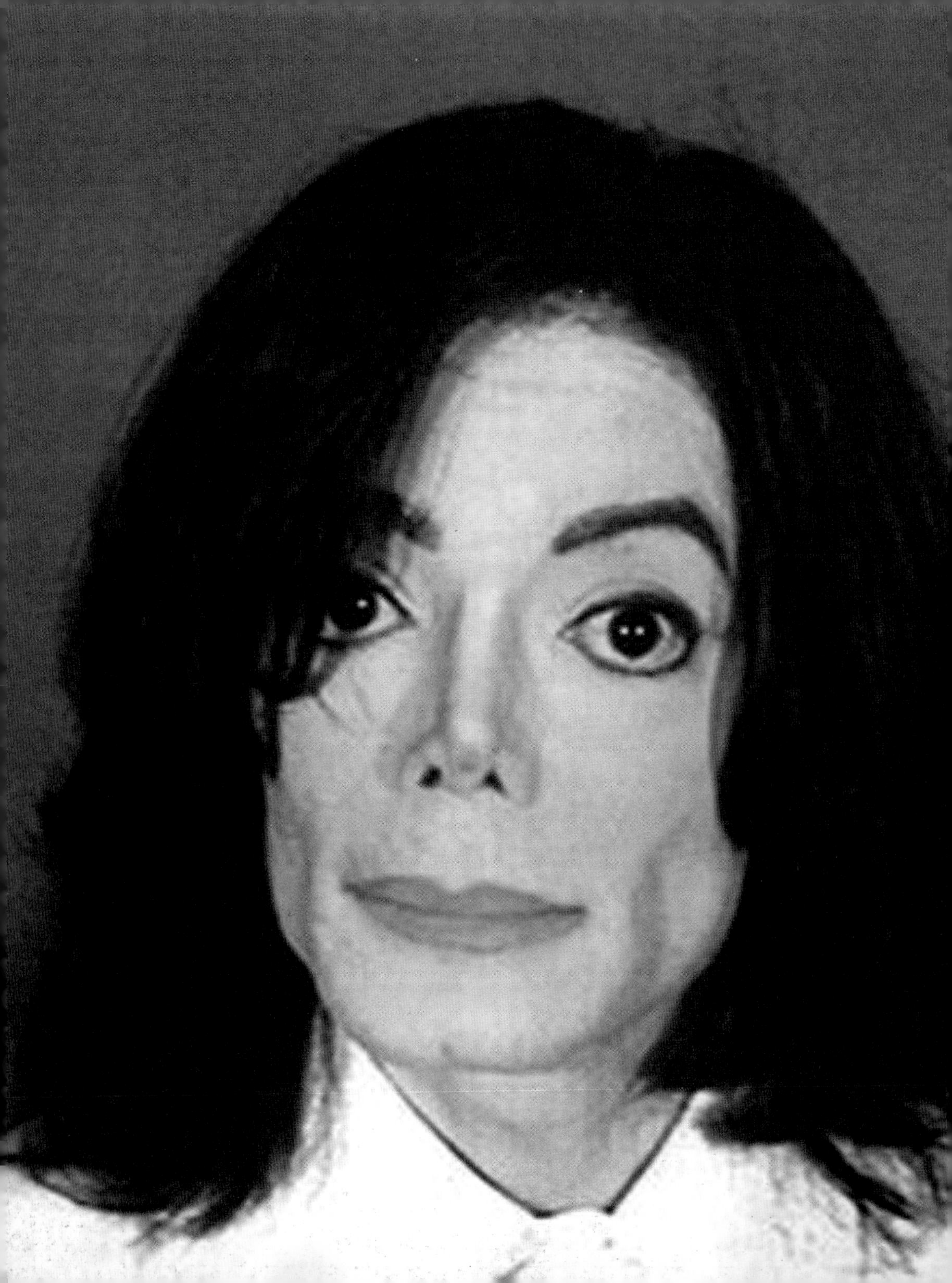

cultural power. Jackson's was the first high profile face to portray a physical transition from African to Caucasian, changing colour and shape – a metamorphosis mirrored in his video morphing 'Black or White'.

The second face is an extraordinary mask, made in New York in the early years of the twentieth century by W. T. Benda. It is made out of laminated paper and is eyeless, and yet it appears so human, so alive. Its elusive enchantment seemed to me to epitomise the spirit or life-force we sense in people's faces and sometimes also, mysteriously, in the representation of faces in art. This book is, in part, an investigation into the mystery of that face. I have also wanted to understand why our faces remain so potent and powerful today despite both our ability to surgically enhance and change them and the transformations made possible by new digital technologies – superseding, some would say, both painted and photographic portraiture.

Thinking about faces through making and representing them has been at the heart of my project. New modes of thought and expression are made possible through the visual, and the conjunction of art and science in which the visual has been a form of knowledge in and of itself (rather than merely illustrative) has always characterised the history of investigations into the face.

As the surgeon Sir Charles Bell explains in his anatomical textbook *Essays on the Anatomy of Expression in Painting*, published in 1806:

> I have often found it necessary to take the aid of the pencil, in slight marginal illustrations, in order to express what I despaired of making intelligent by the use of language merely; as in speaking of the forms of the head, or the operations of the muscles of the face.

Future Face is illustrated with visual material directly referred to in the text and additional images which expand on the themes of the book. Short 'image narratives' on pages 200–211 contextualise this material.

Throughout *Future Face* I have also supplemented my own expertise in literature and the visual arts with material from the fields of physiognomy, psychology, anatomy, medicine and advanced imaging and digital technologies.

Vicki Bruce, a psychologist who specialises in human face recognition, and Alf Linney, a medical physicist who has pioneered computer software for surgery on the face, have each contributed a chapter to the book.

My first chapter, 'Image', looks briefly at the representations of the face throughout history and touches on the long intellectual union of art and science and their similar methods and working practices for the depiction and analysis of the face. The second chapter,

'Bare essentials', explores the physical structure, functions and mechanisms of the face. It examines facial repair, reconstruction and enhancement and their impact on facial appearance and the perception of identity. The third chapter, 'Data face' focuses on reading and interpreting the face, and looks at theories of facial identification and classification. It considers the boundaries between 'real' and artificial faces, how 'new' faces become acceptable, how technologically-produced representations of the face have enhanced or diminished our understanding of the face, and how key ethical issues arise from these and from other recent psychological and bio-medical facial research. The fourth chapter, 'Extreme face' looks at the ways in which monstrosity has been used to understand, interrogate and comment on the face.

Alf Linney's 'Medicine face' revisits some of the material in the second chapter on facial surgery, exploring recent new developments and, in particular, the impact of computer technologies on this process. 'Identikit face', by Vicki Bruce, looks at the material in the third chapter from her own perspective as a psychologist writing about facial recognition and perception with particular respect to current forms of criminal identification, and asks whether machines can enhance this process. And in my Conclusion, I suggest that the relationship between face and identity is neither simple nor direct but remains compelling, and that many of the questions about the face, explored and debated from Aristotle onwards, are still unanswered. As the photographer Philippe Halsman once said:

This fascination with the human face has never left me. Every face I see seems to hide – and sometimes, fleetingly to reveal – the mystery of another human being. Capturing this revelation became the goal and passion of my life.

A. Kertész, **The Puppy**, 1928 (p. 14)
Jo Longhurst, **Terence**, 2003 (p. 15)

face to face

sandrakemp

It never occurred to me before how many faces there are. There are multitudes of people, but many more faces, because each person has several of them. There are people who wear the same faces for years; naturally it wears out, gets dirty, splits at the seams, stretches like gloves worn during a long journey. They are thrifty, uncomplicated people; they never change it, never even have it cleaned. It's good enough, they say, and who can convince them of the contrary? Of course, since they have several faces, you might wonder what they do with the other ones. They keep them in storage. Their children will wear them. But sometimes it also happens that their dogs go out wearing them. And why not? A face is a face.

Other people change faces incredibly fast, put on one after another, and wear them out. At first, they think they have an unlimited supply; but when they are barely forty years old they come to their last one. There is, to be sure, something tragic about this. They are not accustomed to taking care of faces; their last one is worn through in a week, has holes in it, is in many places as thin as paper, and then, little by little, the lining shows through, the non-face, and they walk around with that.

Rainer Maria Rilke, *The Notebooks of Malte Laurids Brigge*[1]

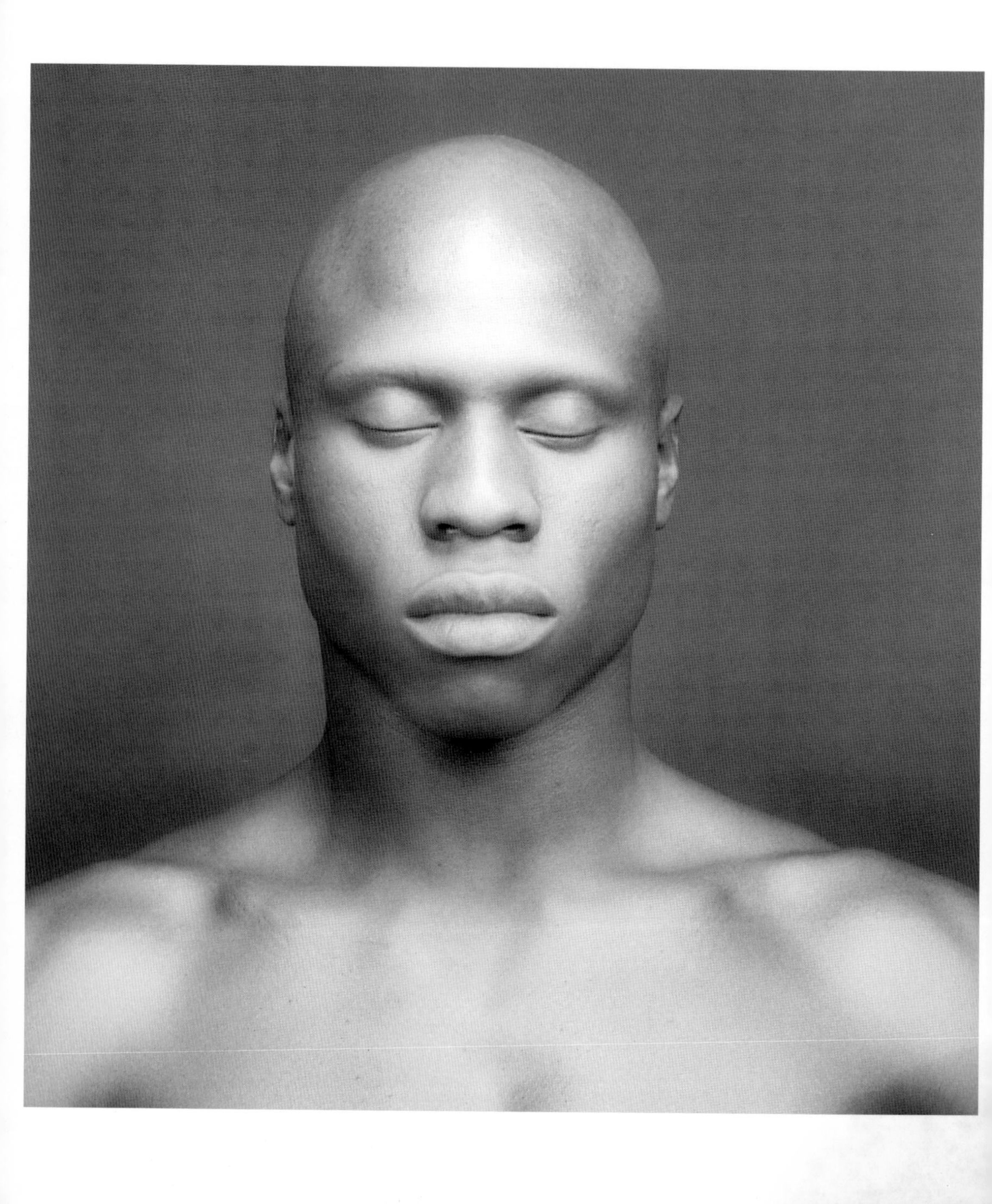

image

We all have a fascination, even a preoccupation, with faces. Openly or covertly, anxiously or eagerly, we catch our reflections in mirrors, glass doors and shop-fronts and 'put on a face to meet the faces' when we go out. Each of the six billion faces in the world is unique, and it is estimated that each face can make thousands of discrete expressions. Babies are genetically programmed to look at faces from moments after birth and children's drawings of stick people with gigantic heads reflect the enormous importance faces play in childhood and throughout life.[2]

Robert Mapplethorpe, **Ken Moody**, 1983

There is beauty and significance in every face. In crowds on streets, in buses and trains and aeroplanes, everywhere, in fact, we see many unremarkable faces. Yet, if we look closely enough, each individual face is compelling. Even when a face is anonymous or vacant, it can engage us.

The face has a strong social, cultural and personal role, is common to all people throughout time, and is therefore extremely engaging. The face is our interface with the world and it is upon the face that our first

Gunter von Hagens, **Plastinated vascular system of the head**, 2000 (p. 22) www.bodyworlds.com

Marc Quinn, **Self**, 1991 (p. 23)

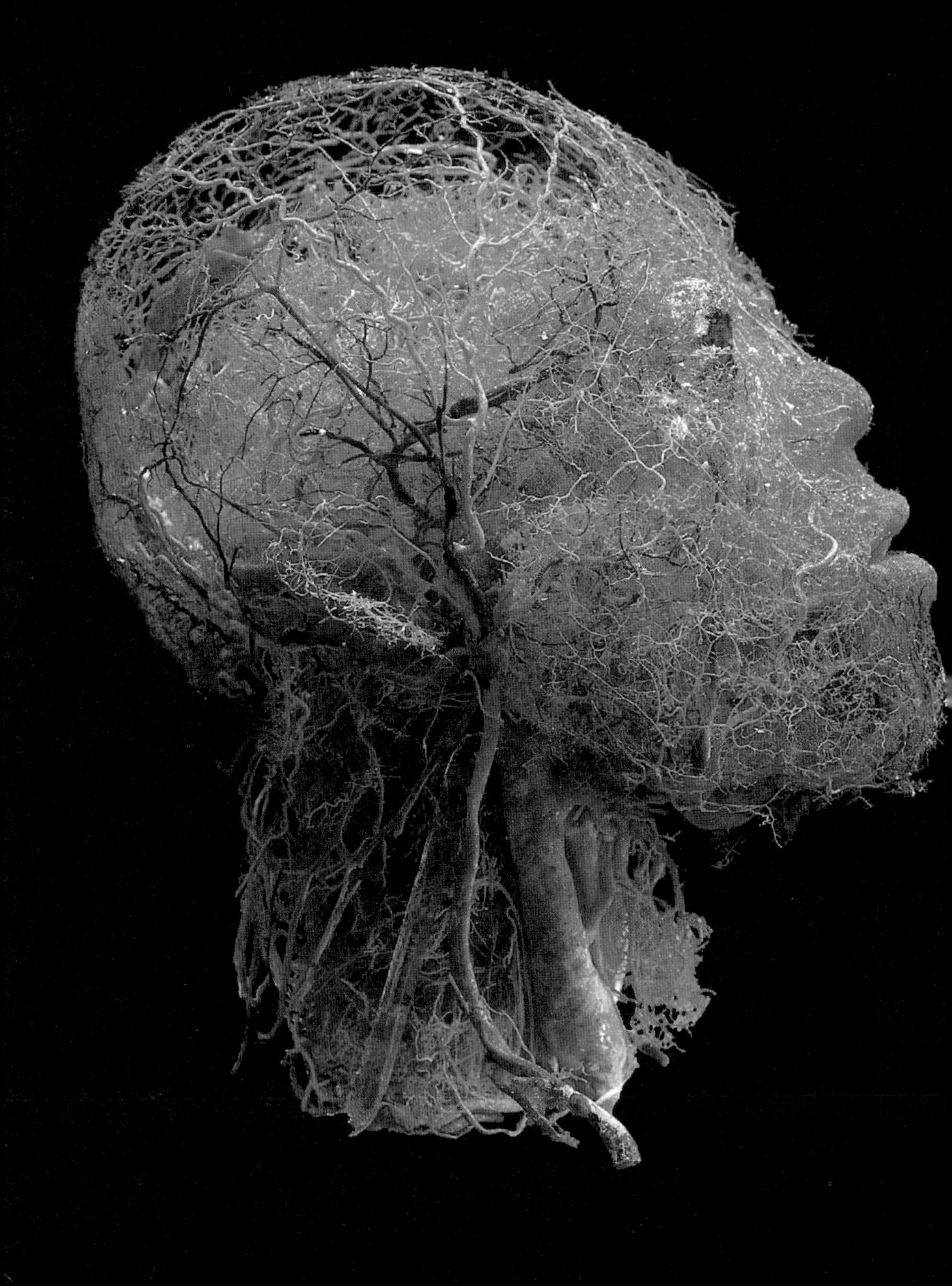

impressions of other people depend. A significant proportion of human culture is based on the reflected visibility of the face. Portraiture is one of the great defining metaphors of Western culture. The art historian Marcia Pointon argues that the history of modern self-identity and subjectivity is inseparable from the portrayal of the human face: 'Modernism and the Portrait might be said to be impacted, welded together.'[3]

History is populated by images of faces. According to legend, Narcissus fell in love with his own reflection in a lake and Saint Veronica found the image of Christ's face miraculously imprinted on a cloth with which she had wiped his face as he carried his cross to Calvary. Representations of faces on skulls in ancient civilisations catch our attention in museums – as do more recent examples of body art, like Marc Quinn's sculptural self-portrait fashioned from nine pints of his own blood. Similarly, Anthony Noel-Kelly's notorious gilt-sprayed cast head of an elderly man caught in the grimace of rigor mortis, and Gunter von Hagens's 'Body Worlds' exhibition, composed of plastinated corpses, both attract and repel.

From Ancient Egyptian mummy portraits to the works in the National Portrait Gallery's annual 'BP Portrait Award' today, the importance of depicting the face retains enormous cultural significance and currency. Despite predictions about the death of portraiture with the turn to abstraction in art at the beginning of the twentieth century, the disappearance of the face was short-lived. New visual priorities, generated by the worlds of media and advertising, cut across the worlds of 'high' and 'low' art and culture with new commercial portrait subjects. Andy Warhol's pop portraits of the 1960s, such as his huge *Marilyn Diptych*, in which 50 identical images of the actress-celebrity, slightly modified in colour and tone, are shown together in a single work, epitomise the celebrity face, as do Mario Testino's contemporary celebrity portraits. In 2004, two exhibitions shown in London are testimony to the lasting currency, complexity and subtlety of the form. The French photographer Philippe Bazin's *Nés* (2000), a series of photographic portraits of new-born babies in a Maubeuge maternity ward, are resonant of a new mode of photographic portraiture. And the exhibition at London's Wallace Collection of twenty-four self-portraits by Lucien Freud, one of the most remarkable painters of the twentieth century, shows the artist experimenting with painted portraiture between the ages of 18 and 80.

Today, images and representations of faces are everywhere, from family photo-albums to police mug-shots, where, as photographer Allan Sekula points out, there is a linkage between the spheres of culture and social regulation: 'every proper portrait has its lurking inverse identification in the files of the police'.[4] In George Orwell's dystopian novel,

Nineteen Eighty-four, social control is characterised by a world which is faceless; in it, O'Brien, the hero's relentless nemesis, says, 'If you want a picture of the future, imagine a boot stamping on a human face – for ever.'[5]

portrait

From ancient civilisations to the present day there has always been something unsettling – even ghostly – about renderings or representations of the face. In the past, portrait painters have been associated with necromancy or trafficking with the dead. For example, the fear that the spirit or soul would be trapped in the reproduction and taken away is common in cultures exposed to photography for the first time. The literary theorist Sarah Kofman has compared the portrait to the uncanny double – to the ghost hovering in a liminal zone, neither living nor dead, neither absent nor present. The belief that the soul or spirit of the deceased will continue to animate the portrait after the originating sitter or subject has died is another lingering superstition. From the sixteenth century onwards, this relation between life and art has been the subject of many representations. Vincent Van Gogh even aspired to make portraits that, a century later 'might appear to the people of the time like apparitions'. Perhaps the best-known literary examples are Edgar Allan Poe's 'The Oval Portrait' (1842) and Oscar Wilde's *The Picture of Dorian Gray* (1891). In Gogol's story 'The Portrait' (1923), horror at the literally deadly rivalry between sitter and portrait is succinctly expressed by the main protagonist, Chertov:

Does the highest art bring a man up to the line beyond which he captures what cannot be created by human effort, and snatches something living from the life animating his model?[6]

The earliest known representations of faces are those from the Graveltian period of prehistory, such as the Brassempuoy Venus, dating from c.23,000 BC. Later, portrait heads were literally constructed from both the dead body of a person and from artistic materials; these 'faces' were built up in plaster over skulls from Neolithic Jericho dating from 7000–6000 BC, with shells set into the eye sockets to simulate eyes. The Ancient Egyptians saw portrait statues as a kind of extension of personality to take the subject into the next world, *simulacra* that the soul could inhabit when even the mummified body had decayed.[7] The apotropaic (evil-averting) image of the gorgon face on Archaic Greek temple pediments had a formal kinship with the *'imagines Clipeatae'* or ancestor faces on shields in Roman houses. Roman portraits were also placed in tombs and had an exemplary role: they were intended to instil in younger members of the family group the virtues practised by their ancestors.[8]

Edwin Romanzo Elmer,
Mourning picture, 1890 (p. 26)

Father and mother with dead child, circa 1850–60 (p. 27)

Rudolf Schaefer, from the series **Dead faces**, 1986 (p. 28)

Professor Dr H. Killian, photograph from
Facies dolorosa, 1967 (p. 29)

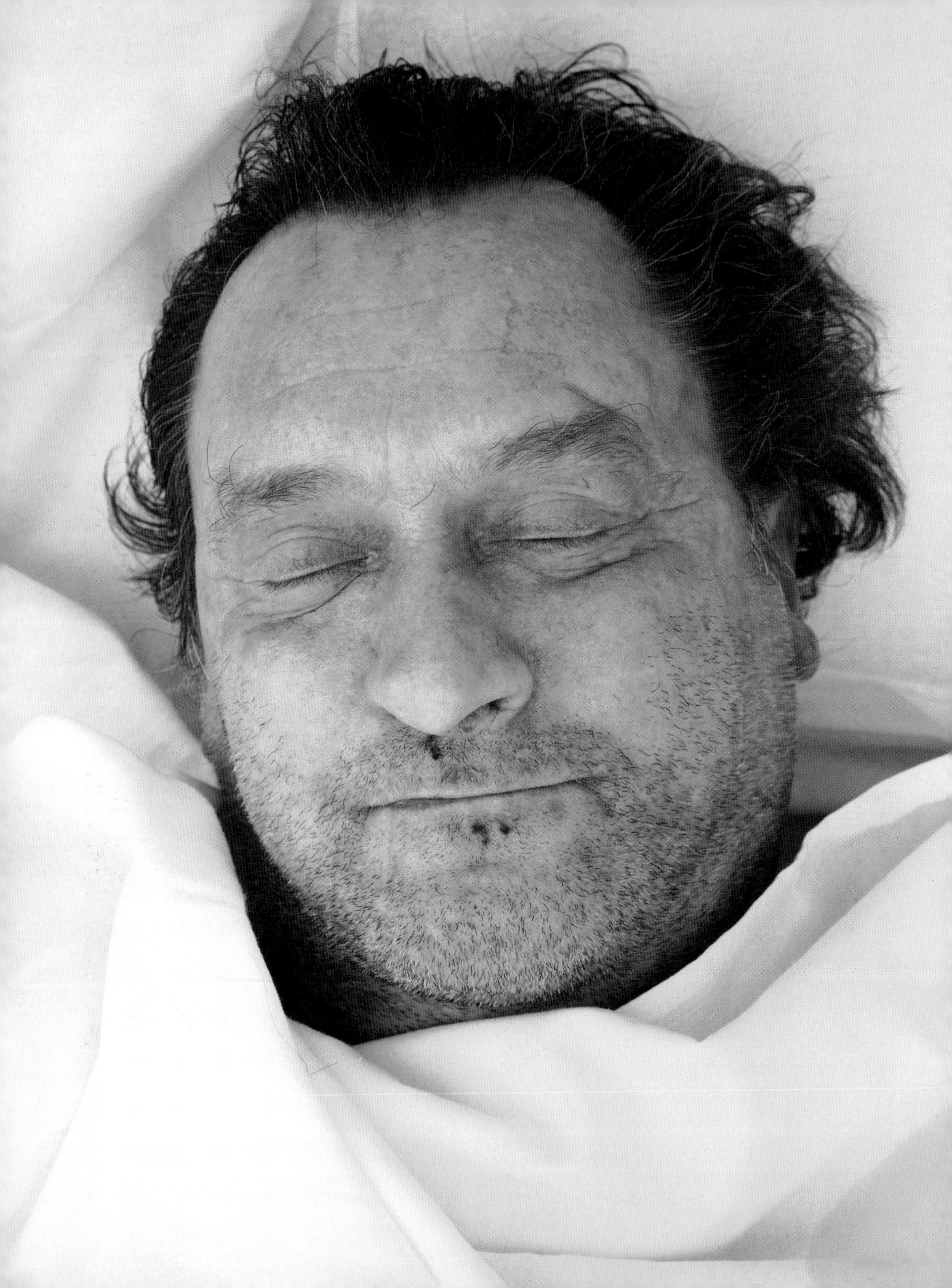

People in the ancient world also sculpted famous faces on busts and statues, minted coins and medallions with faces on them and painted portraits of key public figures on wall paintings and sarcophagi. Even today, famous historical figures, such as Julius Caesar, are recognisable through lasting portrait images on coins, gems and sculpted heads.

In the Hellenistic period, there were the Greek writer Theophrastus's book *Characters* and the Roman Marcus Terentius Varro's 700 painted portraits of particular individuals from the worlds of sport, politics and the academy. As critics Peter Hamilton and Roger Hargreaves note in their book on celebrity portraiture, there is something iconic about these, in the manner of celebrity images today.[9]

Portraiture as we now know it, however, established itself in the dynastic courts of Renaissance Europe as a result of royal and noble patronage.[10] Before that, faces in portraits were generally universalised to epitomise and exemplify a type. Medieval portraits of kings, queens and saints, for example, are symbolic. In a lecture entitled 'The Director faces his collection', Marc Pachter, Director of the Smithsonian National Portrait Gallery in Washington, remarks that the history of art is full of faces that represent anything but particular individuals. The face was often the least important part of the depiction. The aim was to portray a wider philosophical meaning and not a likeness:

In the era of saints' lives and the icons that represent them, as well as in the portrayal of royals, character and greatness adhered least in the particular face. In fact the face was universalised beyond the individual facial quirks to the universal ideals represented by the life. Inquiring whether the icon looked like the saint would have been the most ridiculous question to ask of it.[11]

But in the sixteenth century, Renaissance families established private picture galleries for their portraits, which explained the family's importance and incidentally reflected those conventions that defined its character (by age, gender, beauty, occupation and class). The popularity of portraiture and with it the social construction of the individual face (rather than sophisticated emblematic representations) quickly spread through the upper classes as an expression of power, wealth and social status. By the eighteenth century, the new moneyed middle classes (especially lawyers, the clergy and merchants) were also extending their patronage. From Holbein and Velazquez to Reynolds and Rembrandt, from Degas and Sargent right up to Lucien Freud today, artists have tested and extended their skills and made their fortunes representing the face. Portraiture predominated at the annual programme of public exhibitions of the Royal Academy in London from its founding in 1768, and found a permanent home with the establishment of the National Portrait Gallery in 1856. In support of the Gallery's foundation, Lord Palmerston, the Prime Minister of the day, informed Parliament:

There cannot, I feel convinced, be a greater incentive to mental exertion, to noble actions, to good conduct on the part of the living, than for them to see before them the features of those who have done good things that are worthy of our admiration, and whose examples we are more induced to imitate when they are brought before us in the visible and tangible shape of portraits.[12]

Even today portrait galleries around the world remain concerned with the persona (politician, sportsman, scientist and so on) rather than with the private individual.

By the nineteenth century, however, the development of a more individualist 'humanist' cultural tradition led to a movement towards the depiction of the individual and the idiosyncratic in portraiture. In the expanding visual cultures of commerce and science, there was an increasing desire to find and display the quintessential character of the subject in the face. Portraits began to invite speculation about the personality of the sitter and their inner self and soul. Photography (in the form of 'daguerreotypes' and 'cartes de visite') provided the answer to a widespread and increasingly affordable desire to represent the self and to have a realistic picture of a friend or loved one as a token of them in their absence. As the critic Roland Barthes notes in his *Reflections on Photography*:

It seems that in Latin, 'photograph' would be said 'imago lucis opera expressa'; which is to say: image revealed, 'extracted', 'mounted', 'expressed' by the action of light. And if Photography belonged to a world with some residual sensitivity to myth, we should exult over the richness of the symbol: the loved body is immortalised by the mediation of a precious metal, silver (monument and luxury); to which we might add the notion that this metal, like all the metals of Alchemy, is alive.[13]

The impact of early photography must also be measured in the context of the fact that, until the mid-nineteenth century (when mirrors became commonplace), many people did not really know what they looked like. Not surprisingly therefore, within two decades of the invention of the camera, photography had become an industry, 'a household word and a household want', as the writer and critic Elizabeth Rigby (Lady Eastlake) wrote in 1857, 'used alike by art and science, by love, business and justice'.

The development of cartes de visite in the 1860s created a kind of revolution in portraiture. These low-cost miniature portraits, containing three or four poses fitted to a single glass negative and supplied in bulk, were exchanged and circulated throughout society. They also answered a deeply felt need for pictures of the dead, as a *memento mori*. 'Secure the Substance – Ere the Shadow Fade' was one of the earliest advertising slogans. Extraordinary though it may seem today, from the 1850s until the first decade of the twentieth century, professional photographers openly advertised their willingness 'to take

Brassempuoy Venus (p. 32)
Fruitstone amulet (p. 32)

Plastered skull from Jericho, circa 7000–6000 BC (p. 33)

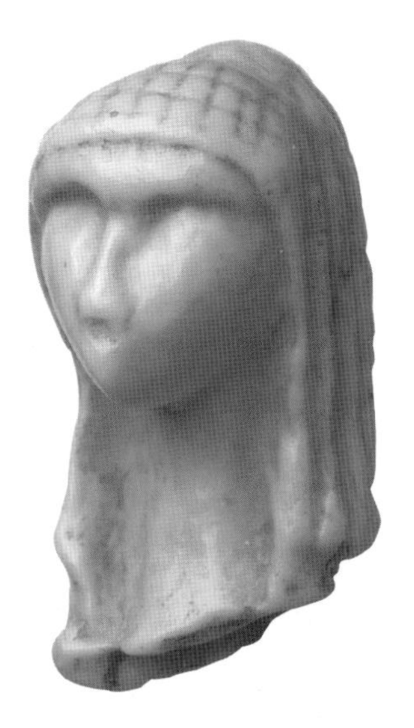

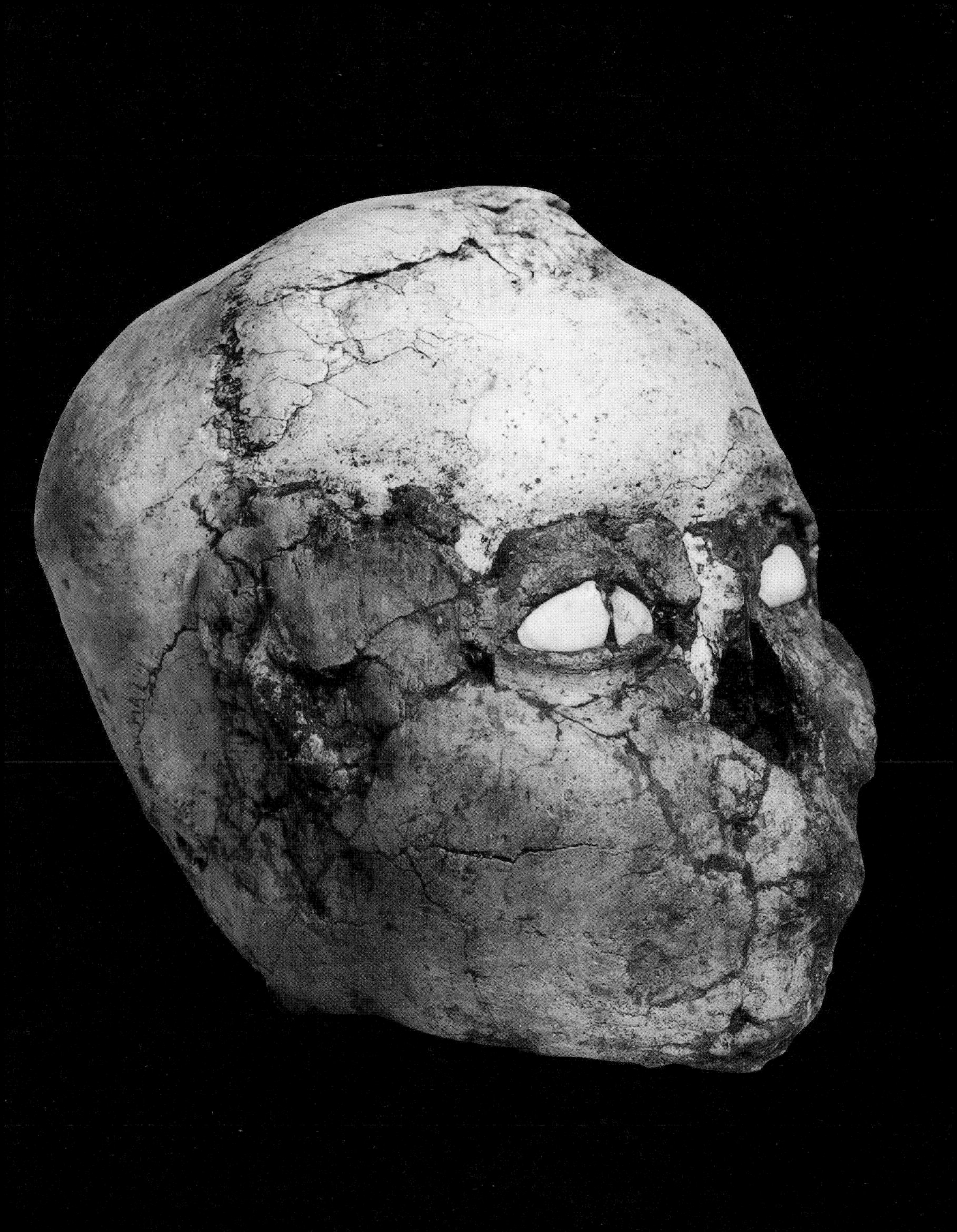

likenesses of deceased persons'. Post-mortem or memorial photographs as they were called, usually concentrated on the facial features of the deceased and were mostly taken in the home. Many photographic studios were almost entirely dependent on this aspect of their trade. It was an age of generally high mortality and family members were frequently apart for long periods. In the 1890s, child mortality rates were particularly high: around one child in five died in the first year of life, and the average was two out of five by the second year. And there is evidence of how desperate families were to preserve something of their dead child. In her book on the subject, the critic Jay Ruby reproduces a *carte de visite* of a baby supported by an adult hand. On the back of the image someone has written 'Taken while dying'.[14] In 1843 the poet Elizabeth Barrett Browning wrote to a friend:

My dearest Miss Mitford, do you know anything about that wonderful invention of the day, called the Daguerreotype?... It is not merely the likeness which is precious in such cases – but the association, and the sense of nearness involved in the thing, the fact of the very shadow of the person lying fixed there forever!... I would rather have such a memorial of one I dearly loved, than the noblest artist's work ever produced.[15]

By contrast, in the first decades of the twentieth century the effect of death on portraiture manifested itself in the temporary disappearance of the face and a movement away from keepsake images or representations of the face of it. After the First World War, the art historian Ernest Gombrich forecast that: 'There will be no portrait left of modern man because he has lost face and is turning back to the jungle.' Klee's abstract painting *A New Face* (of an unnamed subject) (1932) is a characteristic response of this movement towards abstraction, displacement and idiosyncrasy in the early years of the century. In 1901 Klee had recorded in his diaries:

Thought about the art of portraiture. Some will not recognise the truthfulness of my mirror. Let them remember that I am not here to reflect the surface (this can be done by a photographic plate) but must penetrate inside. My mirror probes down to the heart. I write words on the forehead and round the corners of the mouth. My human faces are truer than real ones.[16]

Like Picasso, Braque, Cézanne and others, Klee challenged the relationship between outer and inner, consciousness and countenance. The self is now depicted in decentred and inchoate form. It is evasive, fragmentary and unresolved. Now the photographic mirror of the face is no longer considered adequate or apt. In *Invitation for the Max Ernst Exhibition, Paris, 1935* the Surrealist artist Max Ernst also challenges the conventional notions of a facial reflection of the self in a photo-collage in which the collagist has cut his own face into pieces. The face looks out at the viewer reflected back through the fragments of a broken mirror.

The period of Modernist fragmentation was short-lived, however. Today representations of images of the face remain part of the currency of everyday life. We find them on stamps, coins and banknotes, and on passports, identity cards, bus passes and library tickets. Pictures of writers and celebrities are used in advertisements and on book covers. Caricatures and cartoons are as popular as ever. Portraits appear among boardroom paintings, drawings by street artists, and of course family snapshots. The sentimental currency of the snapshot (now digitised) has been universalised as the way we 'record' our lives. Faces are also used for product endorsement and for manipulating appearances and popular values in the worlds of media and advertising. The ubiquitous smiling and usually female face that stares out at us from the covers of men's and women's magazines (and inside them) is constructed to intrigue and attract – to draw us to the magazine. It is as much a type as some of the examples from classical antiquity. Now the face is of iconic rather than of symbolic importance, advertising a label, a brand or merely a celebrity him (or her) self. Commercial culture has had a profound influence on the face. Sustained media exposure and the development of computer technologies have shown us that the faces we see in newspapers and magazines, in films and on television, have been idealised, manipulated and touched up. The private person is less important than, and often invisible behind, the public image offered as a commodity for the viewer's consumption. In Kundera's novel *Immortality* the protagonist Agnes comments on excessive exposure to faces in the media:

> She looked through the magazine again, from first page to last. She counted ninety-two photographs showing nothing but a face; forty-one photographs of a face plus a figure; ninety faces in twenty-three group photographs and only eleven photographs in which people played a secondary role or were entirely absent. Altogether the magazine contained two hundred and twenty-three faces.[17]

imaging

The question of what a face is and what it is for is intriguing and important for both artists and scientists, and involves integration of the graphic, the forensic, the psychological, the medical and the technological. The face is obsessively compelling. The impulse behind our interest is the urge both to locate an invisible 'self' beneath the skin and to 'read' what the surface appearance tells us. But merely reproducing the face isn't sufficient for this. As Roger Fry once remarked about one of John Singer Sargent's paintings: 'I cannot see the man for the likeness.'

In her 1939 textbook for sculptors, Malvina Hoffman stresses the importance to portraiture of what she calls the 'fourth dimension' or 'psychological atmosphere

Camille Silvy, **Adelina Patti as Martha**, 1861 (p. 36)
Patricia Piccinini, **Psychotourism**, 1994–5 (p. 37)

37_image

of the personality' in adding 'expression and character to the endless variations of mood'. In her 'Suggestions to the student for modeling a portrait', she advocates examining the face both in terms of form (structure, proportion, shape) and as a reflection of character:

The double-sidedness of most faces should be noted, also the difference of the profile as seen from the right or left side and the interpretation of the hieroglyphics that record an inner emotional life. The face must be studied from above, below, three quarters; every fractional move makes a new profile for the artist to observe against the light, and reproduce, if not exactly, at least with as much accuracy as is necessary to reveal the character he is trying to portray.[18]

Both the nature of the information provided in an image and the sensibilities opened up by the image have prompted numerous theories and conventions throughout history about the relationships between appearance and personality. The formal study of 'physiognomy' is said to have begun with Pythagoras, but ancient cultures throughout the world have also studied and read the face.[19] Beliefs and theories about the meaning of the face have led, of course, to stereotypes. These were particularly strong in the nineteenth century, but continue today.

Both artists and scientists have been quick to appropriate modern imaging technologies for their work – reminding us of the long intellectual union of art and science, and how recent the disciplinary divide between them is. From the Renaissance onwards, visualisation has been a way of forming, manipulating, understanding and transmitting knowledge and information. Both artists and scientists sketch, model, scrutinise and make patterns or shape. From photography onwards, new imaging technologies have enhanced development.

Plastic surgeons and artists can now work alongside each other to produce a computer-generated face for the patient that the surgeons then work from during surgery. Surgeons can create 'virtual heads' of patients so that they can plan complex facial operations in order to achieve the desired face with as few scars as possible.

Systems like this can also be used to capture three-dimensional details of living actors to build computer-generated versions and even put dead film stars into new roles. The same technologies that create virtual clones for the cinema can also re-animate the dead for forensic purposes and enhance biometric and criminal identification databases.

In the last thirty years, there have been rapid advances in the description and analysis of facial structures, textures and movements. In the 21st century, humans are composite biological and technological systems. We have potential for reinvention in multiple physical and virtual selves.[20]

The possibilities that exist for re-fabrication allow us to realise the futuristic fantasies of the past and to project new fantasies, which imagine our adaptation to new and emerging technologies and which have been rapidly taken up and developed by artists. 'Because technology changes the way we live and the way we create, it also changes the way we look,' says multimedia artist Daniel Lee, whose work *Self Portraits I, II, III & IV* morphs the stages of his facial evolution from prehistory to the future:

My eyes shrink as electricity eradicates the need to see in the dark. My brain and forehead enlarge as information expands my mind. My features blend as communication brings cultures closer and closer together – Asian, Caucasian and so on. Only the ears remain the same size because we will never stop needing to listen.

The aspects of the face that signify the human, the unique and the enigmatic are accentuated when it is considered next to technology. The face has therefore become a platform – an effective or 'faithful' digital, robotic or avatar face demonstrating excellence or advance in its field.

Adaptations may be mundane, as in MTV's series *I Want a Famous Face*, in which contestants undergo thousands of pounds' worth of plastic surgery to look like their favourite star, and in which transsexual Michael J. Tito has been given jaw and brow sculpting, hormone treatment, cheek implants, and breast and bottom enlargements in order to look like Jennifer Lopez. Or the future, as envisaged in cyberpunk fiction, may allow altered existences in places where living bodies fuse with machine networks and the economy; a state of existence for which the writer William Gibson has coined the phrase 'Terminal Identity'.

virtual face

It is not just visualising techniques and deployment of the image that link art and science in their study of the face. The methods and working practices of each for the depiction and analysis of the face remain strikingly similar – each works through both intuitive and empirical methods of observation and experimentation.

According to Pliny, the first artist to take plaster casts direct from the face as the basis of portrait sculpture was Lysistratus, in the late fourth century BC. Similar techniques are still routinely used in archaeology as well as in forensic science to reconstruct the faces of the dead. Likewise, hundreds of years before Gunter von Hagens's anatomical plastination of the human body, the eighteenth-century Italian artist and sculptor Ercole Lelli developed a technique of building all the muscles in wax on to the bones of an articulated human skeleton. His exquisite and precise anatomical models can still be seen today in the Anatomy Room of the Palazzo Poggi Museum in Bologna.

French sculptor and digital artist Catherine Ikam's electronically captured and digitised 'virtual' faces of her friends and acquaintances are a kind of anatomical waxworks in 3D digital space. These exquisitely beautiful, mobile, moulded digital faces meet your gaze and trace your movements around them. Like all portraiture, Ikam's 'virtual' faces are only (in the novelist Saki's words) 'a dexterous piece of counterfeit life'. But, autonomous and uncannily lifelike, they raise the same questions of presence and absence, body and image, for audiences today as did the early daguerreotypes ('the very shadow of the person lying there') for their Victorian viewers.

In interview with Louis Fléri and Paul Virilio, Ikam says: 'From the beginning [portraits] have been conceived as ways of thwarting absence and death.' She goes on to describe her own virtual heads as 'prostheses' to access 'new architectures of perception'.[21]

Somewhere between original and copy, Ikam's virtual heads remind us that our faces are thresholds. They make us reflect about the relationships between inner and outer, appearance and identity, and about the point at which 'life' becomes 'artificial'. They recast questions which have been asked across centuries and cultures about our continuing fascination with our faces and their images or representations. The American novelist Nathaniel Hawthorne beautifully conveys something of this mysterious attraction:

Nothing in the whole circle of human vanities takes stronger hold of the imagination than this affair of having a portrait painted. Yet why should it be so? The looking glass, the polished globes of the andirons, the mirror-like water, and all other reflecting surfaces, continually present us with portraits, or rather ghosts, of ourselves which we glance at and straight away forget them. But we forget them only because they vanish. It is the idea of duration – or rather earthly immortality – that gives such a mysterious interest to our portraits.[22]

Charles Bell, **The muscles of the face**, 1824

Daniel Lee, **1949–Year of the Ox** (p. 41)

Catherine Ikam and Louis Fléri, **Alex**, 1995 (p. 42)

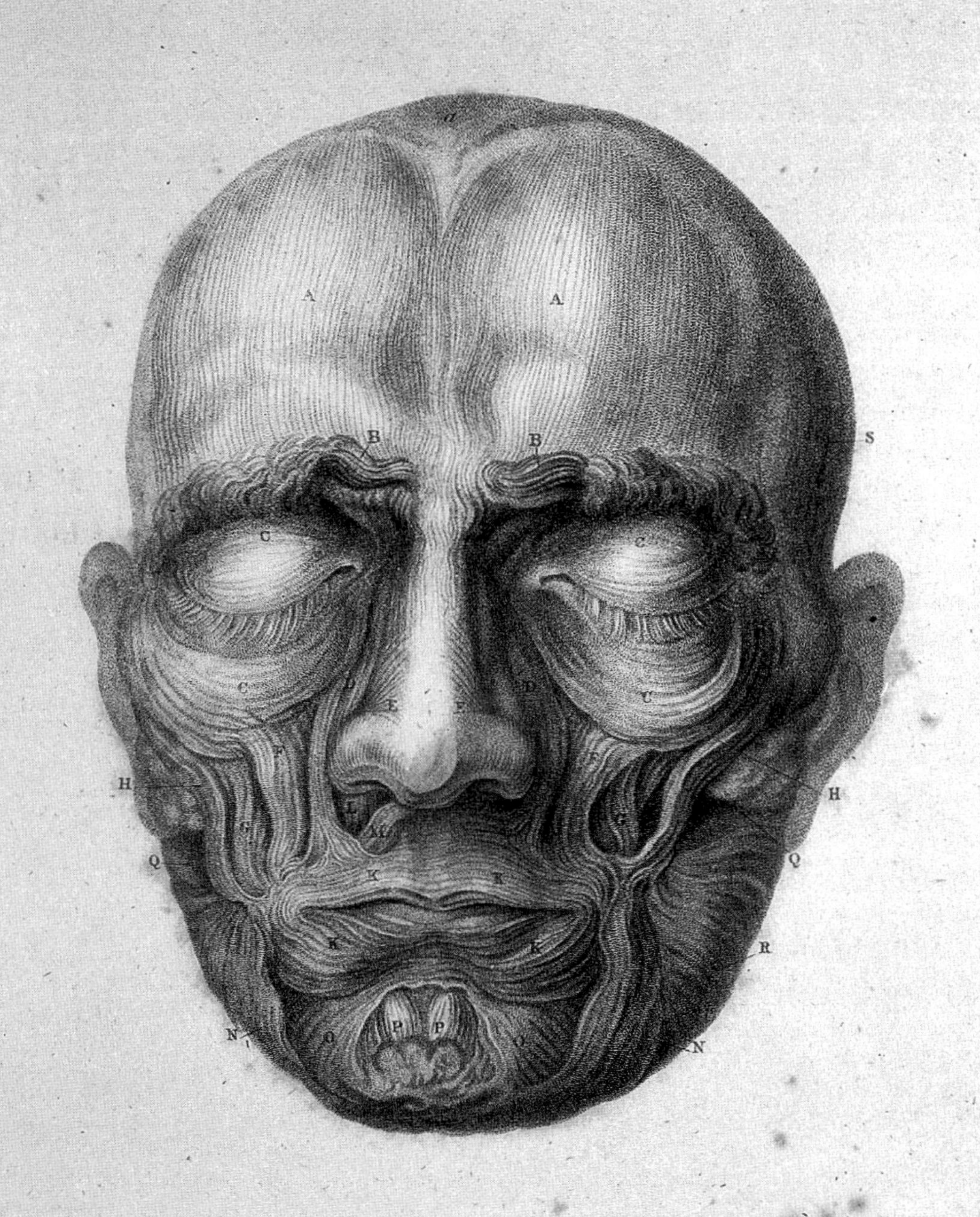

Charles Bell
Engraved by John

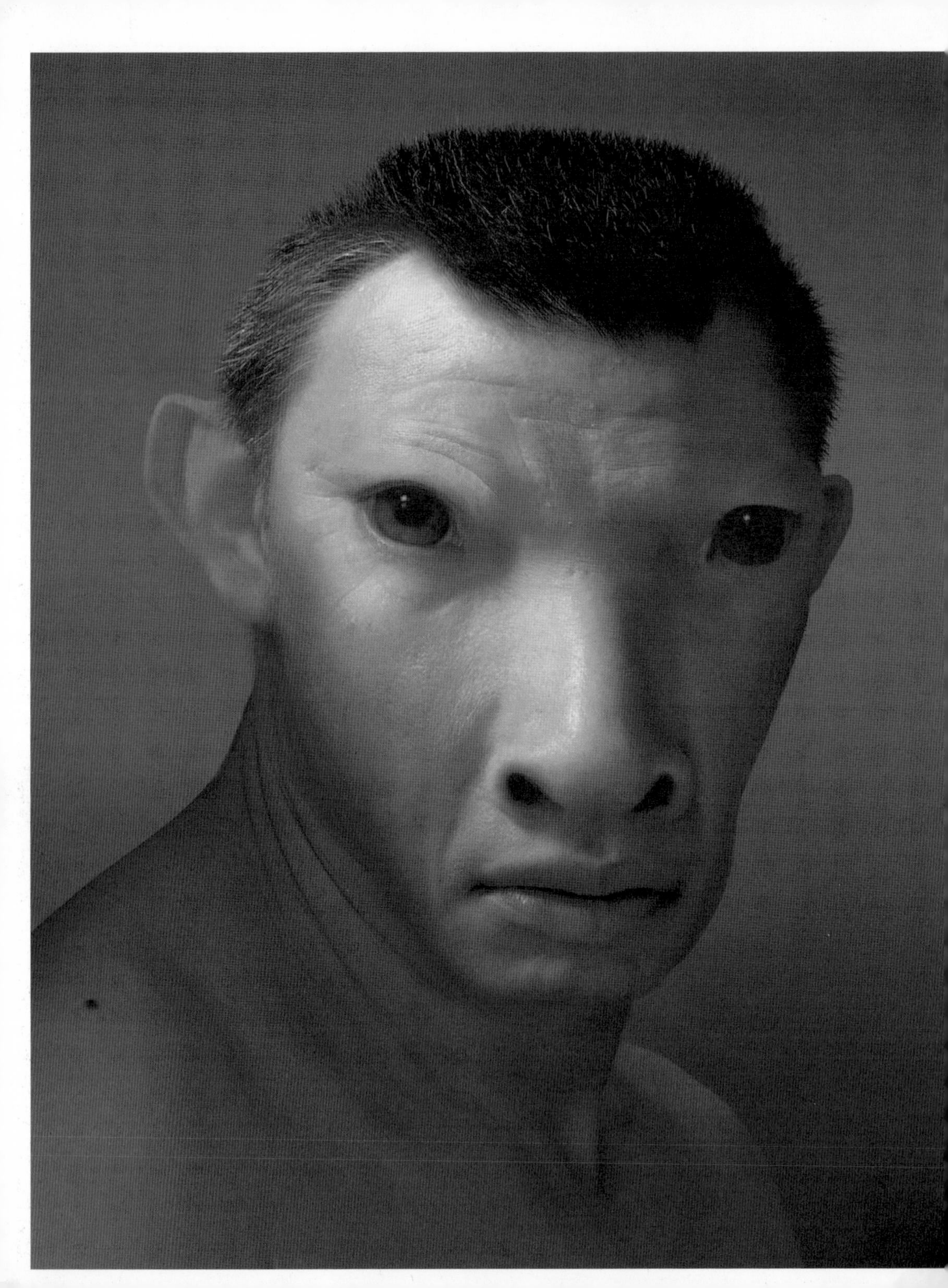

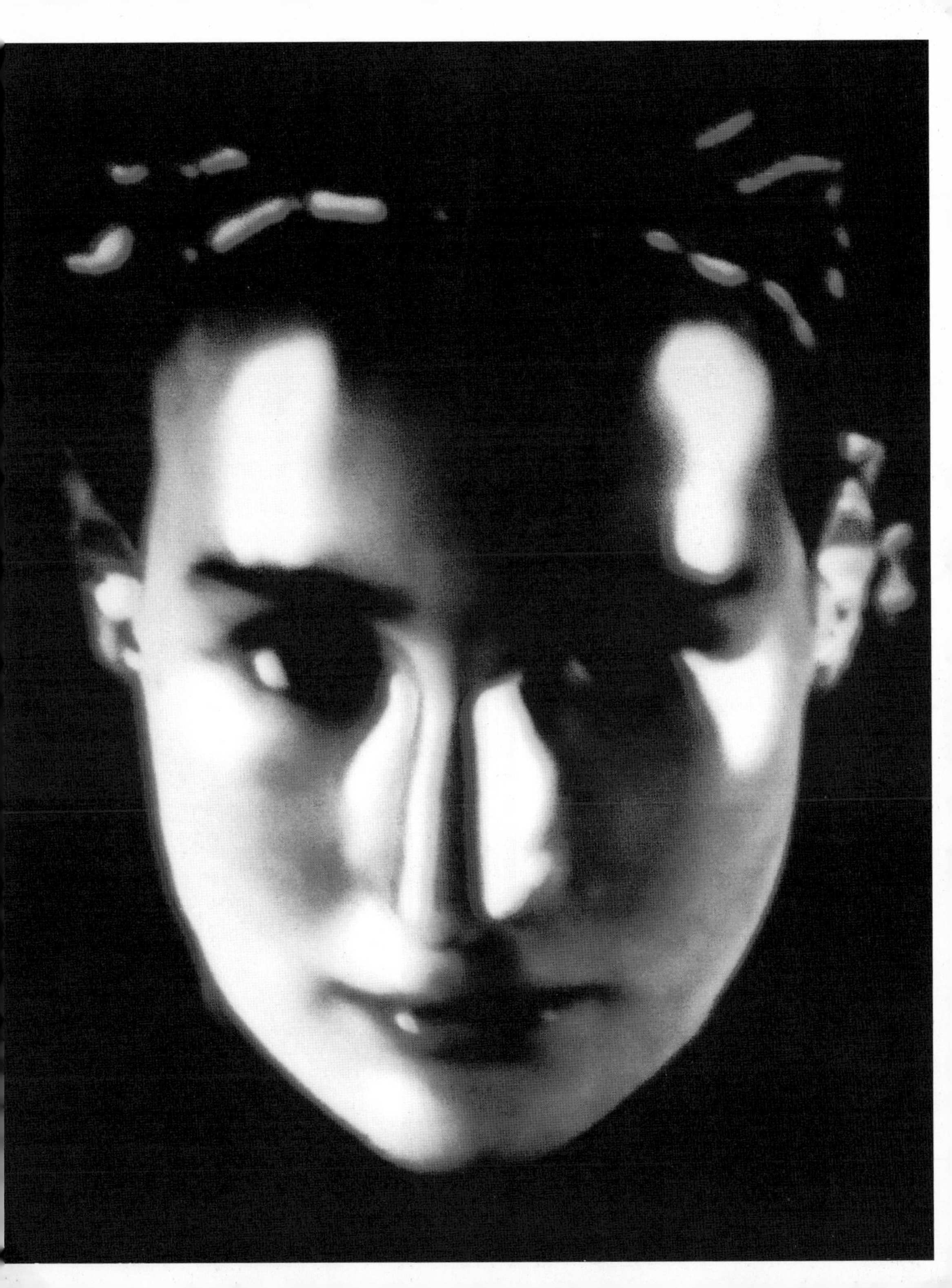

Stripped of all its flesh, the exposed muscle, veins, arteries and nerves wove intricate patterns around each other to provide a latticework of scaffolding behind the skin no longer there to conceal them. Delicate blood vessels, veins and arteries, with tiny tributaries running off them like tree roots, snaked between and beneath bundles of striated muscle spanning the space between her temple and jaw bones like rope bridges . . .

But aside from the obvious usefulness of the arrangement, it made a pattern of breathtaking beauty, though not the human beauty which flits around a complete face as the underlying mechanisms create the mobility we generally recognise as beautiful. Without expression, without even a suggestion of its possibility, this face had the cool beauty of architecture or abstract art, and Bella knew it did not tell a truthful tale about how she might have looked in her fully epidermal form . . . But with all the life subtracted, she had acquired a monumental and timeless symmetry, a still perfection of form which almost stopped the heart.

Jenny Diski, *The Dream Mistress*[1]

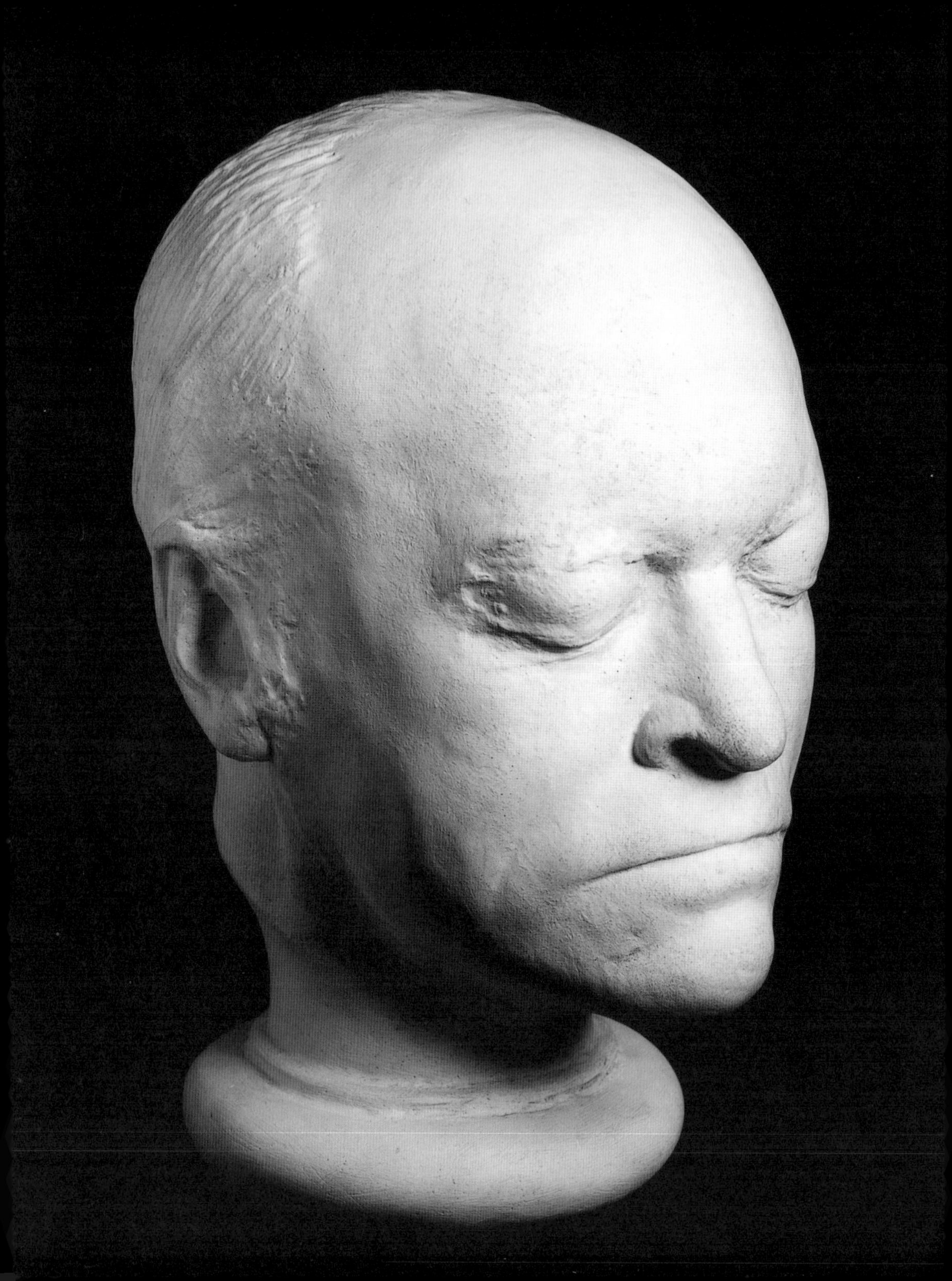

bare essentials

Where does a face begin and end? It may be a skull, a few lines of a cartoon or a caricature, a simple mask, a silhouette. For Jenny Diski's Bella in *The Dream Mistress*, whose face has been removed by laser surgery, 'face' is neither inner nor outer, subject nor object, but exists somewhere between the living human form, science and art.

The structure of the face begins with the facial bones and skull, which facilitate the movement of the jaw and respiration and determine its unique shape (the relative spacing of the eyes, the angle of the jaw and so on).

James Deville, **William Blake**, 1823

Anchored to the skull, and connected and activated by the facial nerves, facial muscles control the movement of the fat and skin that covers them and in which they are embedded. The human eye looks for and can detect the slightest deviation from the accepted norm in another's face.

Facial musculature and its relation to facial expression is still incompletely understood, although this matrix of moving tissue is crucial for the functions of the face (opening and closing the eyes, eating, speaking and so on). The sequence and strength of each

Antonio Durelli, **Ecorché head and neck**, 1837 (p. 48)
Primal Pictures, **Interactive head and neck**, 2003 (p. 49)

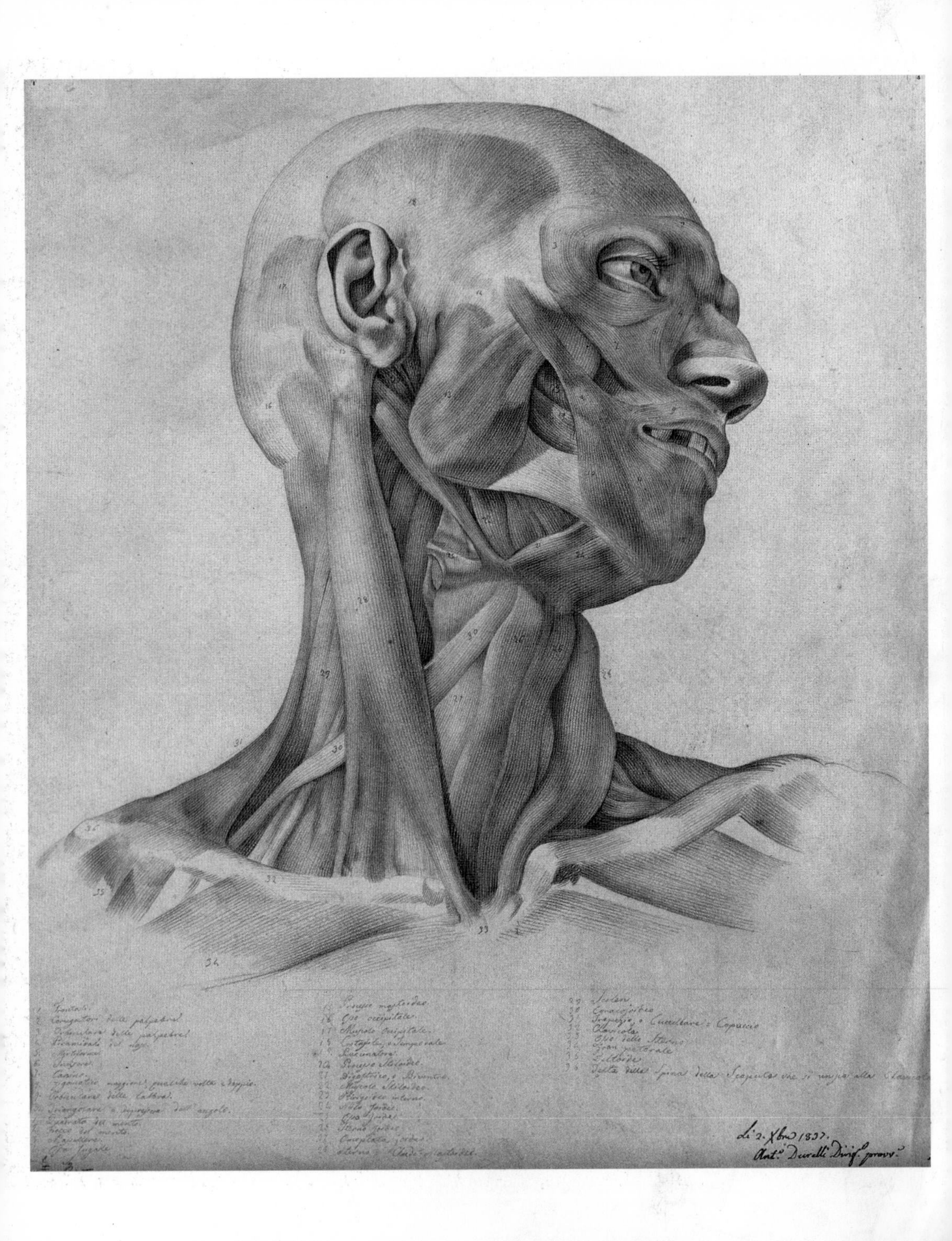

1 Frontale.
2 Corrugatori delle palpebre.
9 Orbiculare delle labbra.
16 Osso occipitale.
17 Muscolo Occipitale.
19 Buccinatore.
22 Muscolo Stiloideo.
30 Coracoioideo
31 Trapezio, o Cucullare, o Cappuccio
32 Clavicola
33 Osso dello Sterno
34 Gran pettorale
35 Deltoide
Li 2. Xbre 1837.
Ant.o Durelli Dis.e provv.o

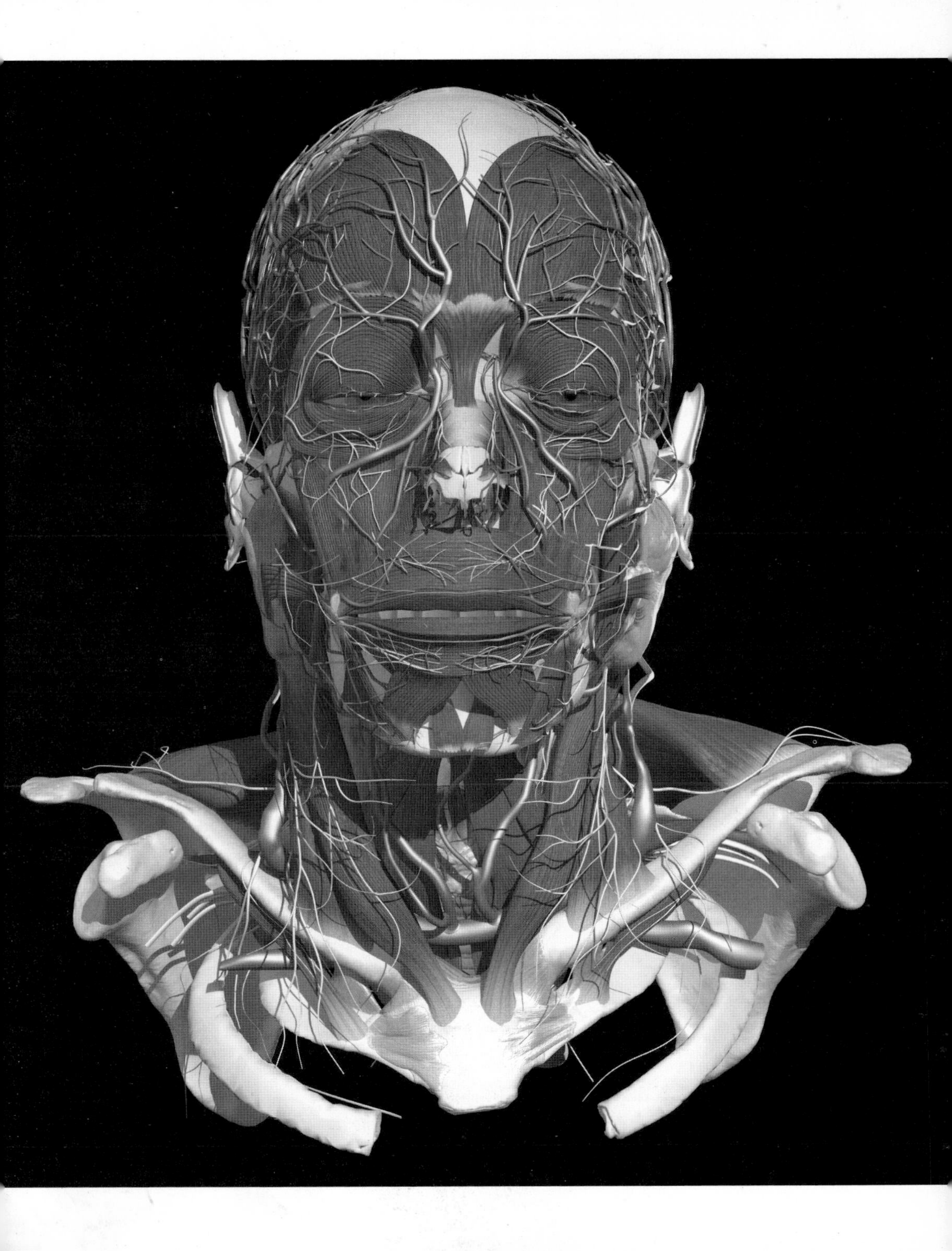

muscle action involved in even the simplest of expressions is quite complex.

In her book *Anatomy for the Artist*, artist and anatomist Sarah Simblet describes how an expert anatomist might claim to tell which language a person speaks from the tone and development of certain facial muscles:

Minute movements of the face tilt the pronunciation of speech and mutate the emphasis of silent communication. Conversely, our first language trains, strengthens and discreetly shapes our instinctive and habitual facial movements. And it is largely the training of the first language that prevents us from mastering the correct pronunciation of a second language.[2]

Facial skin is also highly nuanced. It is more than just a mask covering the muscles and the large amount of fat that lies beneath it. Its colour, lines, wrinkles and general texture are a barometer of a person's age, health and experience. The extraordinary expressive repertoire of the face makes it a compelling subject for both the artist and the scientist.

expression

If the face is central to our understanding of our own identity, facial expressions, both conscious and unconscious, are vital in our encounters and relationships with others. When we communicate in person, we do so through a stream of facial expressions.

Facial expressions depend on complex co-ordinations of nerves and muscles in the face. In 'The Minotaur Syndrome: Plastic Surgery of the Facial Skeleton', surgeon Paolo Morselli describes the surgical procedures performed on a 38-year-old man who had serious social problems because of his aggressive and threatening facial appearance that contrasted with his gentle personality. Morselli coined the term 'the Minotaur Syndrome' to describe the mismatch between the patient's sense of his or her personality and a forbidding facial appearance:

The Minotaur was not a wicked individual, and if it had not been for his facial appearance he would have had a normal life. He was therefore a predestined victim of his facial appearance.[3]

In conversation with the art critic David Sylvester, Francis Bacon argues that portrait painting is 'so fascinating and so difficult' because of the distinction between 'the appearance' and 'the energy within the appearance':

(portrait sitters) prefer a sort of photograph of themselves instead of thinking of having themselves really trapped and caught... if you are doing a portrait you have to record the face. But with their face you have to try and trap the energy that emanates from them.

Describing his own obsession with the smile, Bacon told Sylvester: 'I've always wanted and never quite succeeded in painting the smile,'[4]

Francis Bacon, **Study for a portrait II (after the life mask of William Blake)**, 1955 (p. 52)

J. F. Gautier d'Agoty, **Two heads with brains exposed**, 1748 (p. 53)

and, again, 'I've always hoped in a sense to be able to paint the mouth as Monet painted the sunset.' Bacon himself was strongly influenced by Van Gogh, whose portrait of his doctor Paul Gachet in 1890 does not aim to achieve likeness through 'photographic resemblance' but by means of 'impassioned expressions'.

For forensic scientists, one of the main imponderables is how to reflect the expression on a person's face. A new generation of 'cyber-detectives' use digitised optical imaging to wrap cyberflesh around 3D images of the skull to recreate the face. But, as the work of forensic artist Richard Neave's team of medical artists at the University of Manchester demonstrates, the plastic skills of the sculptor with an intimate knowledge of the workings of human anatomy remain fundamental. The intention is not to create a portrait, but to ignite the spark of recognition that leads to the identification of a previously unidentified body. Richard Neave has verified his technique by reconstructing the face of a living person without knowing in advance who he was. But the technique raises fascinating questions. Professor Peter Vanezis, forensic pathologist and Director of the Human Identification Centre of the University of Glasgow, says, 'You have a man's skull, you create a face, but what expression to give him? If the guy was miserable all his life and you put a smile on his face, no one would recognise him.' For this reason, Richard Neave and John Cragg, Keeper of Archaeology at The Manchester Museum, call their facial reconstructions of historical figures (including Robert the Bruce, an Iron Age Briton and an Egyptian mummy) '3D reports on the skull' to emphasise that these are not portraits, but extrapolations from the bones beneath. The impact is remarkable: the faces draw a strong emotional response from the viewer.

Even with today's technology, facial expression (a mere reflex for humans) remains difficult to simulate robotically. At the NASA jet propulsion laboratory in Pasadena, Texas, the objective is to create a generation of robots that use artificial muscles as easily and comfortably as do humans, both for commercial use and to advance medical technologies. The toy company Hasbro's doll, My Real Baby, released in 2000 and jointly developed with iRobot, a company that began working with Hasbro in 1998, was an early blueprint, combining artificial intelligence, animatronic technology and a web of interactive sensors. The doll was marketed with a promise of fifteen 'human-like emotions', four different stages of laughter and three different stages of excitement. My Real Baby boasted seven different simulated facial muscle groups, was capable of 'hundreds of expressions' from smile to yawn, and had sensors for motion, light and touch. Hasbro claimed that this was a doll 'with the makings of a central nervous system'.

K-Bot, produced in 2003 by David Hanson, a doctoral student at the University of Texas, Dallas, is the face of the future of robotics. It is a mask of elastic polymer placed over

Christian Dorley-Brown, **15 seconds**, 1994–2004 (p. 54)
Richard Neave, **Professor Peter Egyedi**, 1996 (p. 55, left)

Richard Neave, **Reconstructed head of the unknown Dutchman**, 1996 (p. 55, right)

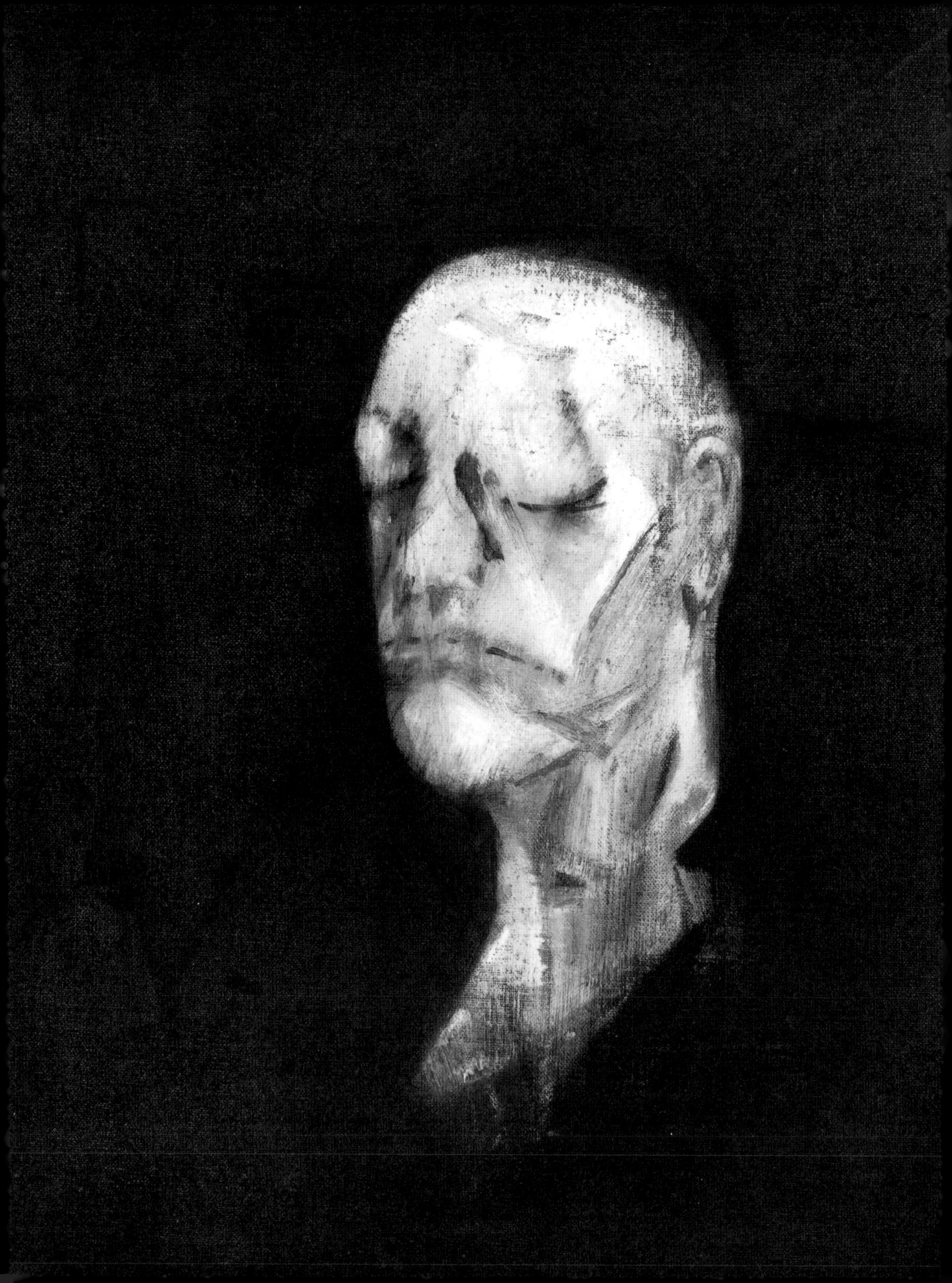

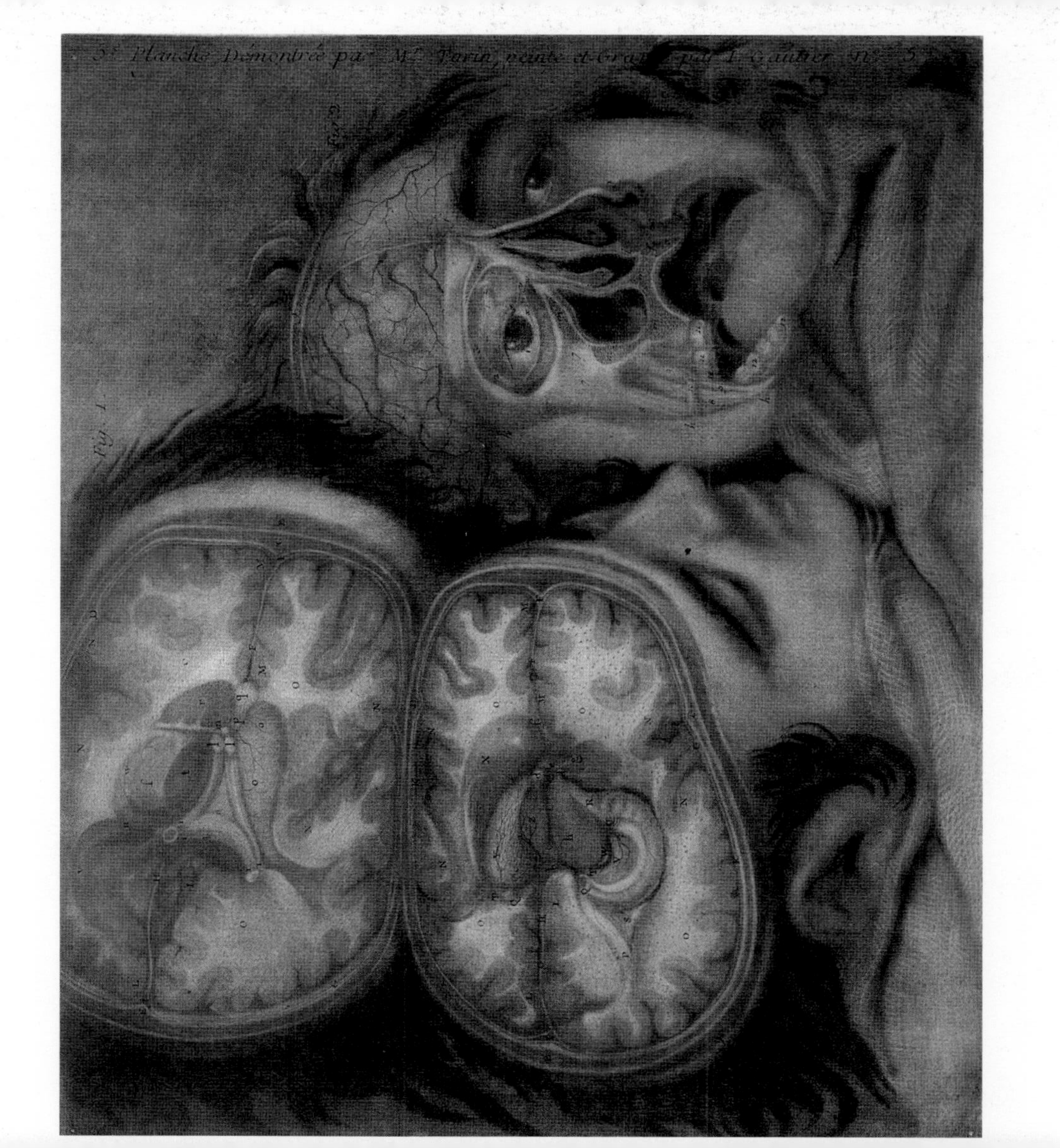
5e Planche Démontrée par Mr Tarin, peinte et par I. Gautier No 5
Fig. 1.
Fig. 2

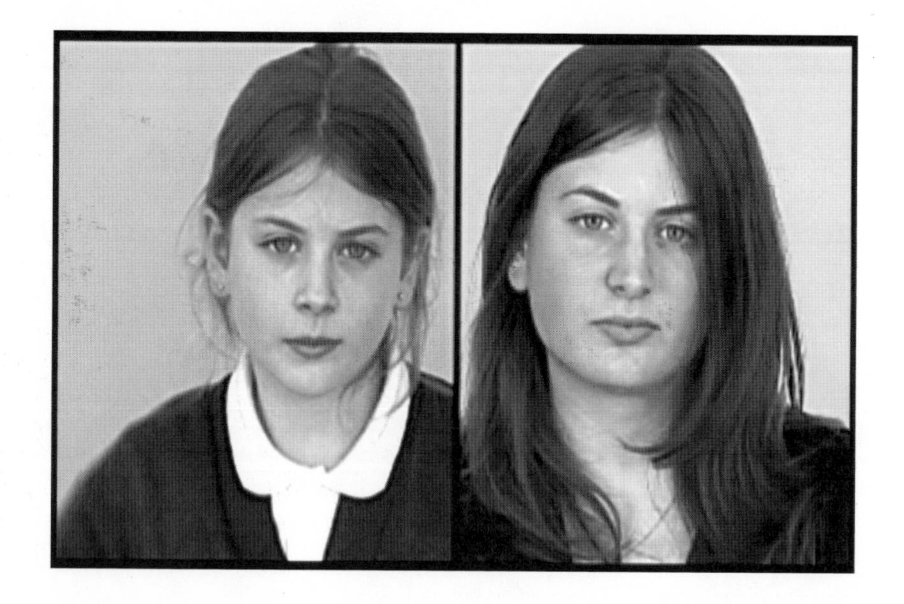

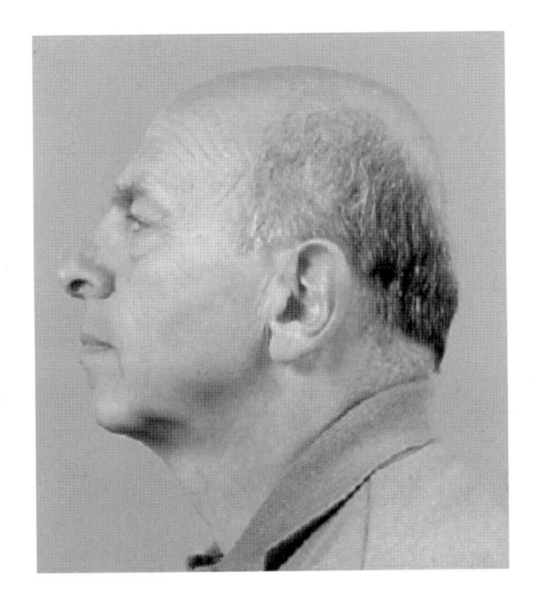

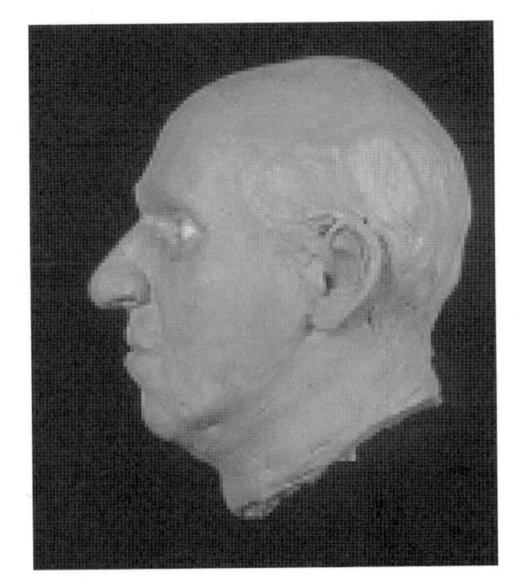

stereoscopic cameras, wiring and tiny motors. It weighs little more than a real human brain and has 24 artificial facial muscles, 28 facial expressions, and eyes that follow you around the room. It is programmed to learn in a natural way, and to recognise and respond to people. It could be of vital medical therapeutic use for people with cognitive dysfunction (like brain damage or autism) and it is a crucial analytic tool for the representation and examination of facial features and expression. Hanson describes his creation as follows:

K-Bot is a human emulation robot, representing a convergence of machine intelligence technology with figurative art. K-Bot makes and sustains eye contact and converses with one in fluent English, while affecting facial expressions as appropriate. In spite of such significant advances, K-Bot is just a preliminary technical sketch ...[5]

expression and personality

If much of our communication of emotions depends on facial expression, does it therefore follow that there is a relationship between facial movement and personality? As early as 1510, Leonardo da Vinci began his anatomical investigations in Milan, and his notebook drawings reveal his examination of the connection between the facial muscles and the expression of emotions. He claimed that 'the good painter has two principal things to paint, that is man and the intention of his mind'. He notes and illustrates the 'muscles of sadness' and 'the muscles of anger'. In around 1660, Charles Le Brun, President of the Académie Royale (the French Royal Academy), gave an influential illustrated lecture series (which was later published), in which he attempted to classify the relationship between emotion and facial expression according to scientific principles. Writing in 1812 in the earliest treatise on the face, the Scottish anatomist Charles Bell noted that 'the thought is to the word as the feeling is to the facial expression'.

But it was Charles Darwin who first suggested that facial expression is a universal language. Darwin published a major study of facial communications in 1872, and illustrated his text with photographic images taken from pioneering neurophysiologist Guillaume Benjamin Amand Duchenne de Boulogne's 1862 work *Mécanisme de la physionomie humaine; ou, Analyse electro-physiologique de l'expression des passions applicable à la pratique des arts plastiques* (The mechanism of human physiognomy; or, the electro-physiological analysis of the expression of the passions, applicable to the practice of the plastic arts). Duchenne linked each fundamental expression to a specific facial muscle or group and tested his theory through the application of localised electric currents administered to the surface of the skin at the point above the motor nerves with wet electrodes to stimulate the muscles. He believed that his work notated a god-given language of facial signs:

Hasbro, **My Real Baby**, 2000 (p. 58)
David Hanson, **K-Bot**, 2003 (p. 59)

(God) wished the characteristic signs of the emotions, even the most fleeting, to be written briefly on man's face. Once this language of facial expression was created, it sufficed for Him to give all human beings the instinctive faculty of always expressing their sentiments by contracting the same muscles. This rendered the language universal and immutable.[6]

The photographer Adrien Tournachon worked with Duchenne in the early stages of the project to produce the series of photographs in *Mécanisme* showing men, women and children whose facial muscles have been artificially distorted into expressions of extreme emotion, or what Duchenne calls 'gymnastics of the soul'.

As a result of his work on expression, Darwin believed that many expressions and their meanings (such as astonishment, shame, fear, horror, pride, hatred, wrath, love, joy, guilt, anxiety, shyness and modesty) are universal: 'I have endeavoured to show in considerable detail that all the chief expressions exhibited by man are the same throughout the world.' Darwin prepared a questionnaire on expression, which he sent out to missionaries, teachers and colonialists in remote parts of the British Empire. The questionnaire asked respondents to note the expressions of aboriginal peoples. From this survey, along with many years of his own observations, Darwin concluded that those expressions he observed in England were the same as those described elsewhere. He then attempted to find out why people with no cultural links vent their emotions in identical ways.

Darwin eventually concluded that expressions are innate rather than learned behaviour. If we had to be *taught* to smile, everyone would do it differently. Smiling, crying and the other emotional expressions thus fall into the category of instinctive behaviour. According to Darwin, the universal facial expressions can be traced back either to our common prehistoric ancestors or to our infancy, when they performed some useful, instinctive function. Even though these expressions have long ago ceased to be of any use, adults continue to perform them through habit whenever the feeling originally associated with the expression arises.

Various forms of theatre have shown how the actor, with supreme control over the facial muscles, can employ the face like a mask to simulate expressions of joy, astonishment, anger. The stock facial expressions and bodily gestures of classical theatre, for example, demonstrate this. During the late nineteenth century, the pantomime character of Pierrot also exemplified the artifice of facial expression. Mime artist Marcel Marceau developed this skill further, and the French Surrealist Claude Cahun's montage *I.O.U. (Self-Pride)* (1930) includes a handwritten message: 'Under this mask, another mask. I will not finish taking off all these faces.' The 'ten thousand and one facial expressions captured in the form of masks' in the Surrealist

Guillaume Benjamin Amand Duchenne de Boulogne/Tournachon; **Mécanisme de la physionomie humaine**, 1862 (p. 60)

Guillaume Benjamin Amand Duchenne de Boulogne **Electrisation**, 1862 (p. 61)

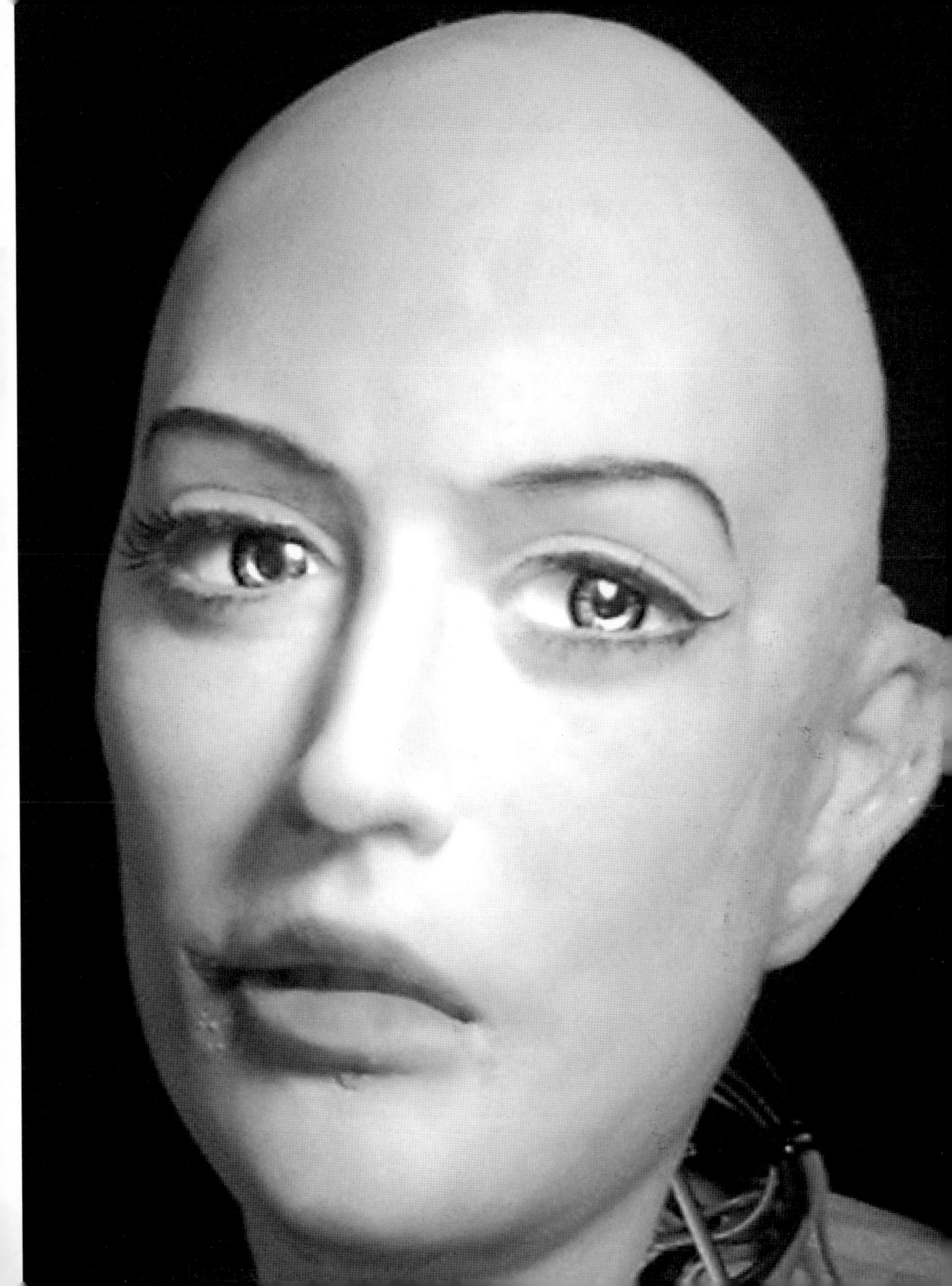

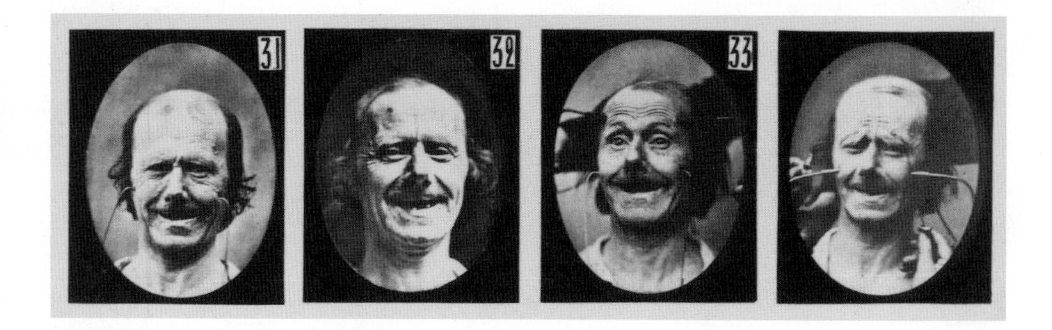
31
32
33

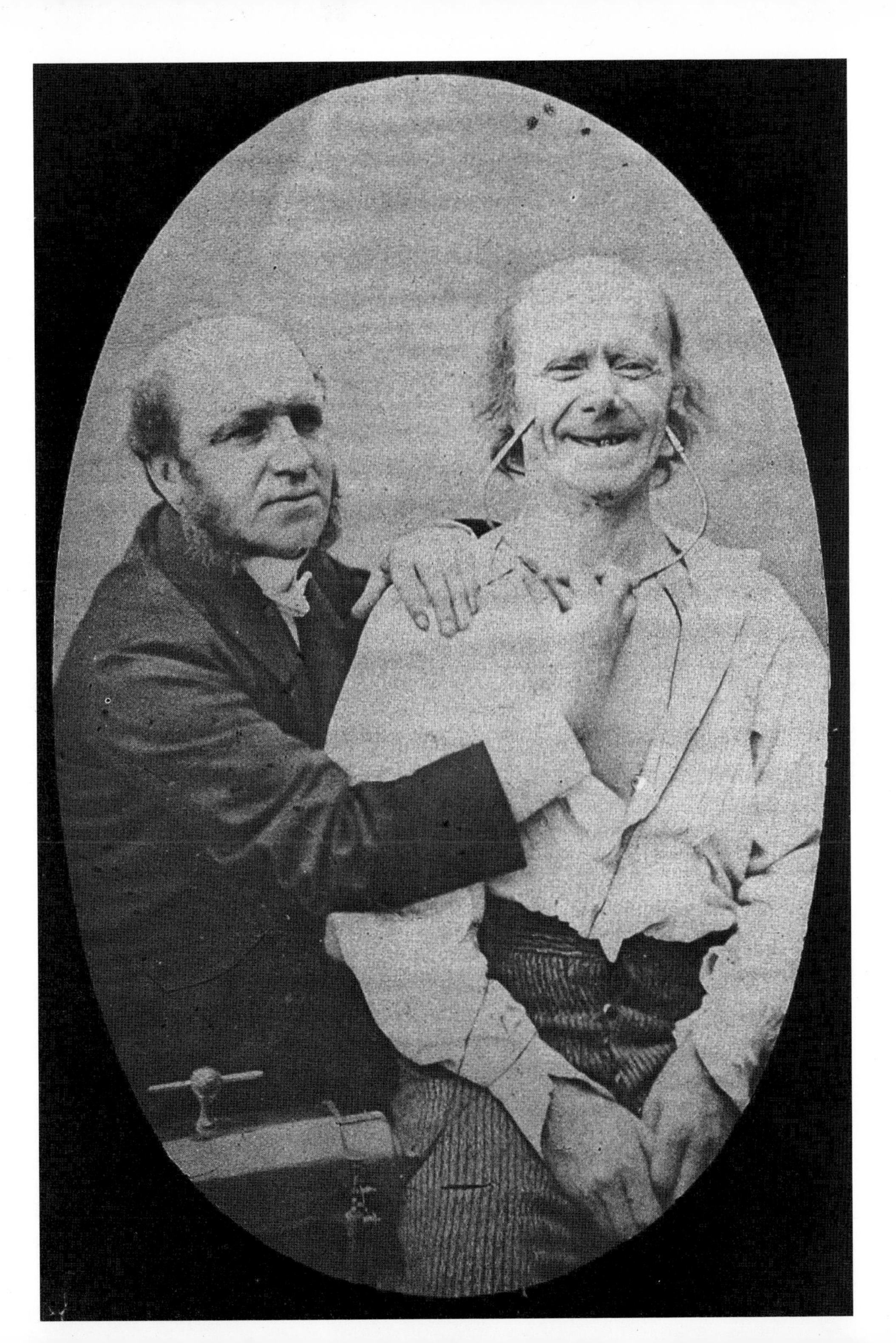

French dramatist Antonin Artaud's Theatre of Cruelty operate independently of the emotions expressed. The management of facial expression is also something we all do in daily life (as for example, when concealing disappointment or distress when we lose at sport or are upset). A number of cross-cultural studies conducted over the past sixty years support Darwin's view that facial expressions are universal and a product of our evolution. Recent research carried out by Paul Ekman, Professor of Psychology at the University of California Medical School, and others claims to have identified seven principal facial expressions. These appear to be recog-nisable across a wide range of different cultures, and are so universal that even children who have been blind and deaf from birth display them. They are: anger; disgust/ contempt; fear; happiness; interest; sadness; and surprise. Ekman also claims to have demonstrated that each individual face can make more than 10,000 expressions. 'As part of that work,' he tells us, 'I had to learn to make every muscle movement on my own face. Sometimes, to verify that the movement I was making was due to a specific muscle, I put a needle through the skin of my face to electrically stimulate and contract the muscle producing an expression.' In 1978, jointly with co-researcher Wally Friesen, Ekman published the Facial Action Coding System (FACS), which is now used extensively to measure facial movements. Since then, Ekman has controversially argued that he has identified the facial signs that betray a lie:

What I have termed *micro expressions*, very fast facial movements lasting less than one-fifth of a second, are one important source of *leakage*, revealing an emotion a person is trying to conceal. A false expression can be betrayed in a number of ways: it is usually very slightly asymmetrical, and it lacks smoothness in the way it flows on and off the face.[7]

Facial expression and its relationship with thought, feeling and consciousness is a crucial subject of enquiry for philosophers. In *Remarks on the Philosophy of Psychology*, Ludwig Wittgenstein describes the human face as a 'picture' of inner emotional states:

We describe a face immediately as sad, radiant, bored, even when we are unable to give any other description of the features. Grief, one would like to say, is personified in the face. This is essential to what we call 'emotion'.

For philosophers Maurice Merleau-Ponty and Emmanuel Levinas, the human face is not just the vehicle for the expression of inner emotions. Merleau-Ponty argues that our personalities or selves are reflected back to us by other people through facial communication: 'I live in the facial expression of the other as I feel him living in mine,' he says. For Levinas, faciality is the 'first philosophy' or 'primordial poem': a visual dialogue that precedes both language and notions of character or personality, and incarnating subjectivity through the act of looking or being looked at.

In this respect, the face is the manifestation of existence, or of being, itself, or, metaphorically, of 'the trace or image of God' in humanity.[8]

Questions about the relationship between face and identity were raised by Jonathan Miller's 1998 exhibition at the National Gallery in London, 'Mirror Image: Jonathan Miller on Reflection', which centred on the representational ambiguities of the self. 'There is something paradoxical about seeing ourselves from the viewpoint of another person,' Miller notes, '... and yet by the time we are two years old, we scarcely give it another thought: and a significant proportion of human culture is based on the reflected visibility of the personal self.'

The same issues have always fascinated and dogged portrait artists: how we perceive ourselves; the relation between artist and sitter, subject and object; the assumptions we make about a person from considering their likeness. As the philosopher Roland Barthes writes in *Camera Lucida*, his posthumously published reflections on photography:

Four image-repertoires intersect here, oppose and distort each other. In front of the lens, I am at the same time the one I think I am, the one I want others to think I am, the one thephotographer thinks I am, and the one he makes use of to exhibit his art.[9]

The clinical neuro-physiologist Jonathan Cole also argues that it is very difficult to know our own faces, even with the help of photography and, more recently, video. Because we usually see our own faces mirror-reversed, they have a more complex and elusive image to us than to other people. Nor can we be certain that our image of ourselves is the same as the one that our friends and loved ones have of us.

In his book *About Face*, Cole interviews people whose faces have altered their lives and writes about the relation between facial mobility, intelligence and understanding: how facial problems may affect social development and selfhood. The subject of his first study, Mary, is a woman who suffered facial paralysis through a stroke, and then wasted away because she could no longer communicate. 'Without a face she as a person was all but invalidated,' Cole remarks. His other subjects also reveal how the immobility of the face affects the experience of emotion itself and how much our well-being depends on an emotional sensibility revealed to others via the face:

The case histories and narratives in this book ... tell of difficulties in the calibration and experience of emotion and in the essential deeply embedded role of the face in our perception of self... All tell of the essential role of the face in the expression and experience of feeling itself. Those who were hardly aware of the facial origin of their problems show how deep within us are these matters that they are only brought to light by a shattering disconnection between personality and the face.[10]

Claude Cahun and Marcel Moore, **MRM (sex)**, circa 1929–30 (p. 64)

Elia Alba, **Doll heads (multiplicities)**, 2001 (p. 65)

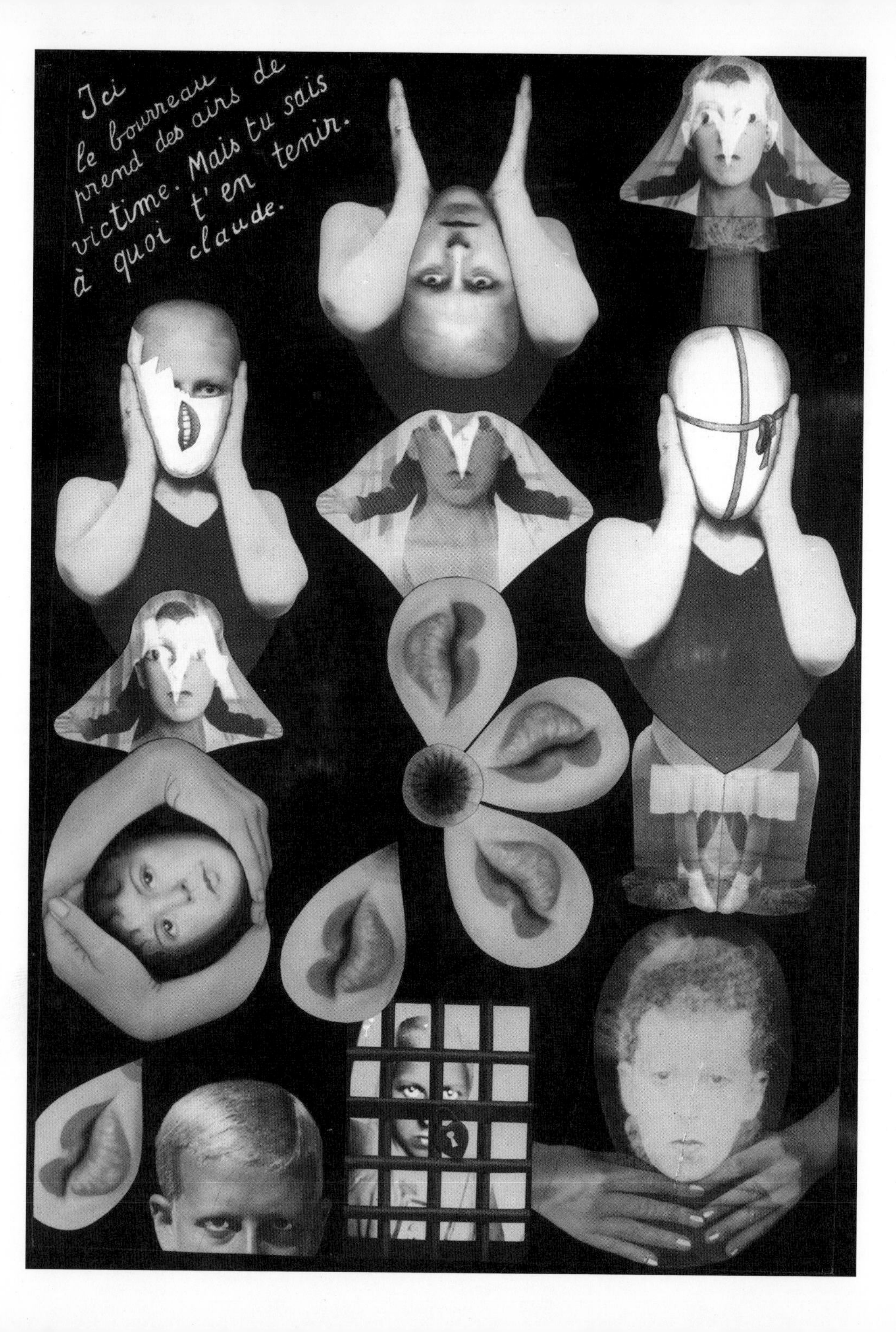
Ici
le bourreau
prend des airs de
victime. Mais tu sais
à quoi t'en tenir.
claude.

Recent research also indicates that the act of forming a facial expression has an impact on how we feel. We now believe this in common experience, but before the research it wasn't an intuitive link. Facial muscles feed information to our brains, so, for example, when the brain recognises that we are smiling, it releases a hormonal response that accompanies a state of happiness. More research is needed before we can understand how mood is affected in those who are unable to form expression in conventional ways. However, there are numerous reports of the difficulties experienced by those who are unable to use their faces to communicate effectively, whether through the absence of expression, or miscommunication resulting from altering expressions.

without expression

Disruption to one's facial appearance, especially the inability to recognise oneself, represents a profound disruption of body image and may constitute a major life crisis. There is rare condition called Moebius Syndrome, named after the late nineteenth-century German neurologist Paul Moebius, which leaves patients incapable of facial expression and movement as they are born without the cranial nerve that operates them. People with Moebius Syndrome are unable to blink or smile and their faces appear like a tightly drawn mask.

In the 2003 Reith Lecture Series entitled 'The Emerging Mind', Vilayanur Ramachandran, Director of the Center for Brain and Cognition at the University of California, drew attention to the significance of other facial syndromes such as prosopagnosia ('face blindness') and Capgras Syndrome (when a patient regards close acquaintances as impostors). Prosopagnosia usually occurs when there is damage to a structure called the fusiform gyrus in the temporal lobes on both sides of the brain. As a result of this damage it becomes impossible to recognise people's faces. Ramachandran describes the case of a patient who had been in a car accident, had sustained a head injury and was in a coma for a fortnight. When the patient emerged from the coma, he was quite intact neurologically, but had one profound and sustained delusion: he would look at his mother and say, 'Doctor, this woman looks exactly like my mother, but she isn't – she is an impostor.' Ramachandran suggests that whilst the visual areas of the brain are completely normal in this patient, the connections to his emotional centres have been damaged in the accident. This means that the patient can recognise his mother visually but not emotionally, leading him to believe that the person he sees is an impostor posing as her.

Because of its association with the 'self' or with 'identity', physical damage to the face has particular powers of horror. Disfigurement predominantly signifies horror – the horror that we feel when we see a distorted face is conflated with the identity that the face betrays. Films like *The Phantom of the Opera* (dir. Rupert

Julian, 1925), *The Hunchback of Notre Dame* (dir. Wallace Worsley, 1923), *Nosferatu* (dir. F. W. Murnau, 1922) and *Frankenstein* (dir. James Whale, 1931) all imply that a monstrous face motivates monstrous behaviour.

Occasionally, however, examples of more humane and rounded films about the relationship between face and identity and the impact of facial disfigurement hit the big screen. *The Elephant Man* (dir.David Lynch,1980), *The English Patient* (dir. Anthony Minghella,1996) and *The Officers' Ward* (dir. François Dupeyron, 2001) are all examples of this. *The Officers' Ward* is a World War I story based on a novel by Marc Dugain which tells the story of a handsome French officer, Adrien E., who spends five years in the specialist Val-de-Grâce hospital near Paris following serious facial damage from the impact of a German shell. The film's use of disfigurement is neither sensationalist nor sentimental. The plot turns on an extra-ordinarily powerful scene in which Adrien sees his rebuilt face for the first time. The camera pans across the ward and a nurse refuses Adrien's request for a mirror to look at his face. Adrien takes off his bandages and looks at his reflection in the window-pane instead.

The Officers' Ward questions what kind of life is possible (and if life is desirable at all) following facial disfigurement. It makes us ask why we fear what might be behind the bandages or surgical mask and what makes damage to the face so horrific and so profoundly disturbing. 'Suicide watch' on new patients who have yet to confront their damaged faces is part of the daily routine in the film. It was also the case at Queen Mary's Hospital, Sidcup, and at the real-life Val-de-Grâce hospital during World War I. In *The English Patient* (based on the novel of the same title by Michael Ondaatje), a badly burned pilot (Ralph Fiennes) is pulled out of the wreckage of his plane in the Sahara desert. Identified only as 'the English patient', he is placed in the care of an army nurse (Juliette Binoche). He is described as 'A man with no face. An ebony pool. All identification consumed in a fire . . . There was nothing to recognise in him.'

rebuilt face

Surgeons have been reconstructing faces for centuries, but as Andrew Bamji, consultant anaesthetist and archivist at the Queen's Hospital, Sidcup, pointed out when speaking at a conference on Art and Surgery at University College London in 2002, 'the Great War of 1914–18 threw up injuries of a type and scale that had never before been a feature on the battlefield.'[11] At the new Queen Mary's Hospital, Bamji's archive of medical records, prostheses, paintings, photographs and textbooks relating to the development of facial surgery during World War I is an extraordinary testimony to the hospital's pioneering surgical procedures. Trench warfare resulted in horrific facial

Henry Tonks, **Studies of facial wounds** (p. 68)
Mark Gilbert, **Chris (i) & (ii)**, 1999 (p. 69)

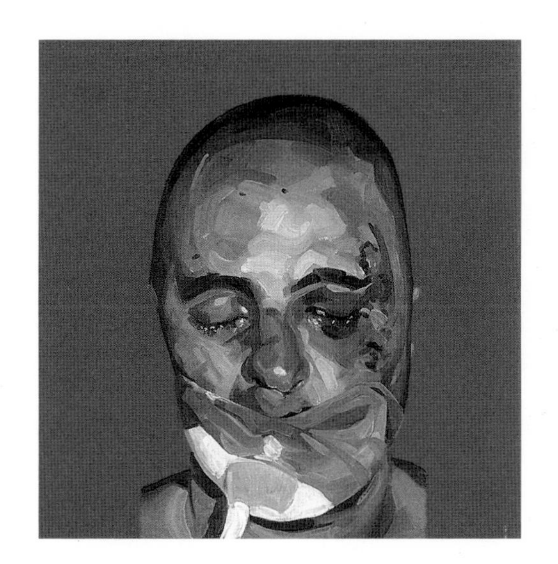

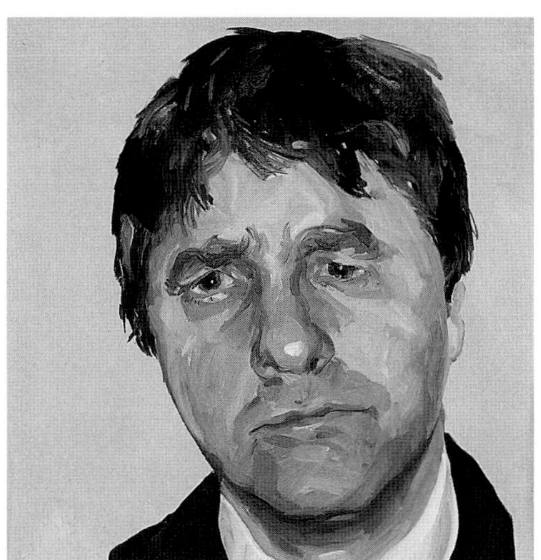

injuries as armies came under heavy continuous shelling in fixed positions, and any soldier who raised his head above the parapet of a trench was in constant danger of being shot. Although steel helmets helped to prevent injuries to the head and increased survival rates, surgeons in the field were suddenly confronted with facial injuries en masse.

For example, on 1 July 1916, the first day of the Somme offensive led to a flood of facial casualties that turned hospitals into what one surgeon described as 'chambers of horrors':

There were wounds far worse than any we had met before... men without half their faces; men burned and maimed to the condition of animals. Day after day, the tragic, grotesque procession disembarked from the hospital ships and made their way towards us.[12]

Two hundred facial injuries were expected in the base hospitals; two thousand arrived.

In France alone, 3 million injured were treated in military hospitals, 500,000 of whom had suffered injuries to the head.

The leading surgeons on the Allied side were the New Zealander Harold Gillies and the Frenchman Hippolyte Morestin. On his return from France to the Cambridge Military Hospital at Aldershot in 1915, Gillies put together a team of dedicated facial surgeons, dentists and anaesthetists, supported by radiologists, medical illustrators and sculptors. The ex-surgeon and illustrator Henry Tonks, later to be Professor of Fine Art at the Slade School of Art, moved from Cambridge to join the team. A new hospital specifically for facial injuries opened in 1916 at Sidcup. This hospital was organised on national lines, with separate contingents of staff from Great Britain, Canada, New Zealand and Australia. The heads of these sections would become pioneers of plastic surgery in their own countries after the war.

Between 1917 and 1925, some 5,000 servicemen were treated at Queen Mary's and its associated hospitals. Drawing on 282 case-notes of New Zealand casualties and over 2,000 British equivalents, Andrew Bamji describes how many surgical lessons were learned at Sidcup. The hospital also became the birthplace of modern anaesthesia, since the then-current method of placing ether masks over the face was obviously unsuited to operations to correct facial injury. A form of intratracheal delivery, using a narrow-bore tube so that ether use could be kept to a minimum, was developed on injured veterans returning from Passchendaele.

Despite extraordinary medical advances, the contemporary attitude to facial injury both before and after surgery was one of horror or disgust. Benches on the road into Sidcup were painted blue to warn local residents that the men sitting on them might have a disturbing appearance. In 1918, in an article entitled 'How the American Red Cross in

London Mends Mutilated Faces', Katherine de Monclos of the American Red Cross explains how the hospital for facial wounds in France was more generally known as 'A museum of horrors'. She notes that the 'pitiful hideousness' of the veterans with wounded faces made them 'difficult to place'. 'Even the public highways are forbidden to them,' she goes on, 'unless they continue wearing a bandage to hide the gaping hole, the absence of a chin, or the loss of a nose.' A compassionate view was expressed by Ward Muir, a journalist who worked as an orderly corporal at the 3rd London General Hospital during the war:

There is one perturbing experience which, for the worker in such an institution, is inevitable. It is this: He finds that he must fraternise with fellow-men at whom he cannot look without the grievous risk of betraying, by his expression, how awful is their appearance. Myself, I confess that this discovery came as a surprise. I had not known before how usual and necessary a thing it is, in human intercourse, to gaze straight at anybody to whom one is speaking, and to gaze with no embarrassment.[13]

Writing about the impact of faces destroyed in the Napoleonic Wars a century before, Carl Ferdinand von Graefe, a German ophthalmologist of the time, reflected social views that are strikingly similar:

We have compassion when we see people on crutches; being crippled does not stop them from being happy and pleasant in society ... (but those) who have suffered a deformation of the face, even if it is partially disguised by a mask, create disgust in our imagination.[14]

How were artists to portray the new phenomenon of what Wyndham Lewis, in his emphatically entitled book *Men Without Art,* called society's 'shell-shocked men and war idiots, its poison-gas morons and its mutilated battle wrecks'? In 1916 and 1917 Henry Tonks made 75 pastel sketches of casualties with severe facial injuries from the Western Front, arising from his work for Harold Gillies' team. These so-called 'Studies of Facial Wounds' are both portraiture and pathological record, through which history is witnessed in a catalogue of wounds upon which each individual's subjectivity appears to be based. Tonks's official role was to draw diagrams of the surgery carried out on the patients, but he also began to draw portraits of the soldiers before and after treatment to supplement the photographic records in their medical files. Each sketch depicts a fusion of art and anatomy, a curious blend of damaged inner structure and outer surface glimpsed through gaping wounds and features lost in fugitive blur.

As the art critic Emma Chambers points out:

In looking at these portraits the viewer constantly tries to make sense of the mismatch between the ruptured exterior casting of skin and the internal layers of flesh; the bodily presence and the interior identity of the sitter.[15]

Tonks called his sketches 'the poor ruined faces of England' and refused to submit them to the Imperial War Museum because he did not wish them to be on public display. In retrospect, they are an extraordinary record of the history of surgery and art.

Previous artists had for the most part emphasised noble and heroic moments of war in their portraits of soldiers rather than facial injury and disfigurement. (Charles Bell, Goya and some American Civil War artists are perhaps notable exceptions.) Most early photographic depictions of facial disfigurement appear stylised beside these deeply moving portraits of ordinary men with neatly combed hair, sharp shirts and ties and damaged faces.

However, in German pacifist Ernst Friedrich's *War Against War* (1924), a picture book documenting the horrors of World War I, the 24 photographs of the rebuilt faces of veterans whose faces had been horrifically damaged in the war shocked Europe. Friedrich used them to represent the 'visage of war'. They are a chilling demonstration of the fact that no more horrible result of war could be represented in the public sphere than the mutilation of the face. Indeed, this result was too horrible to be generally contemplated, and the damaged face rarely appeared as part of the visual memory of the war – the wounded were represented by those with missing limbs but still 'handsome' faces, an iconography that continues in war movies and war literature up to the present day.[16]

Nearly a century later, facial damage remains a sensitive, even taboo, area of fine art. The National Portrait Gallery in London showed two very different exhibitions in 2002: a collection of Mario Testino's glossy billboard-sized fashion and celebrity portraits and, in sharp contrast, 'Saving Faces: Portraits by Mark Gilbert', the result of a highly unusual artist's residency initiated by Ian Hutchison, a surgeon specialising in corrective facial surgery for accident and trauma patients and those with degenerative diseases such as facial cancer. Like Tonks's pastel surgical drawings, Gilbert's work conveys the enduring power of the face to communicate intensely individual personality and emotions, despite having suffered horrific damage and disfigurement. Unlike Tonks's sketches, however, Gilbert's portraits were made for public display, in order to portray the possibilities of modern surgery. Hutchison also hoped that the experience of sitting for portraits would be a cathartic one for his patients, helping them to accept their altered appearance. 'Research into facial and oral cancers have very few advocates and it is a very isolating, very lonely disease. Historically, it has affected men more than women, and men are more reticent and unwilling to talk about their diseases.'

One of the most striking of Gilbert's portraits is that of the barrister, Henry de Lotbiniere, who was operated on sixteen times by Hutchison over fourteen years for a severe facial cancer. All the patients apparently expressed a strong desire to have their portraits exhibited to help educate the public about the nature of facial disfigurement, to challenge the negative associations of facial disfigurement and to reduce the discrimination that many patients feel exists. 'It's great, my picture,' says Chris, who had been attacked by seven men with baseball bats; 'I'm like a Frankenstein by Andy Warhol.' More than a million people in Britain are affected by facial disfigurement each year, but despite significant medical advances not enough is known about facial disease, injury and deformity, or their psychological and emotional impact.

Significantly, facial surgeons have historically been associated with the restitution of physical appearance as opposed to surgery that saves lives. Restoration of the obliterated features revives self-respect and a sense of 'self' and 'identity', and, starting with the work of Ambroise Paré, the sixteenth-century sculptor and father of facial prostheses in the West, sculptors have played a vital part in the development of facial surgery.

Ambroise Paré's fabrication of facial prostheses did much to form the basis of facial reconstruction by prosthetic means. Writing in 1579, he described his procedure as follows:

When the whole nose is cut off from the face ... it cannot be restored or joined again for it is not in men as in plants. Instead of the nose cut away or consumed, it is requisite to substitute another made by art ... this nose so artificially made must be of Gold, Silver, Paper or Linen cloths glued together; it must be so coloured, counterfeited and made both of fashion, figure and bigness, that it may as aptly as possible resemble the natural nose.[17]

In 1832, the 22-year-old Alphonse Louis, a French artillery gunner, was wounded in the trenches during the Siege of Antwerp by a splinter from a 12-inch shell. He became known as the 'Gunner with the Silver Mask' after having a prosthetic mask made by an artist from Antwerp, M. Verchuyler, from the design of a Dr Forget, to compensate for the loss of much of his lower face. Fashionable mustaches and whiskers were used to hide the lines along which the mask met the remaining facial tissues. The mask was also painted in oils to match the natural colour of Louis' face. This exercise was reported in the *London Gazette* as being so successful that even at close quarters it was not evident that he was wearing a prosthesis. It was also reported that as well as restoring the function of the face, the mask collected secretions and supported the tongue, thus making the prosthesis both cosmetic and functional, the two characteristics used to measure its success.

Eighty years later, during World War I, whilst Tonks was sketching facial injuries and Gillies was operating on them, sculptors Francis Derwent Wood in London and Anna Coleman Ladd (who had studied with Wood) in Paris were making similar masks for veterans whose faces had been surgically

Joseph Clover administering chloroform, 1862 (p. 74)
Gunner with the silver mask (p. 75)

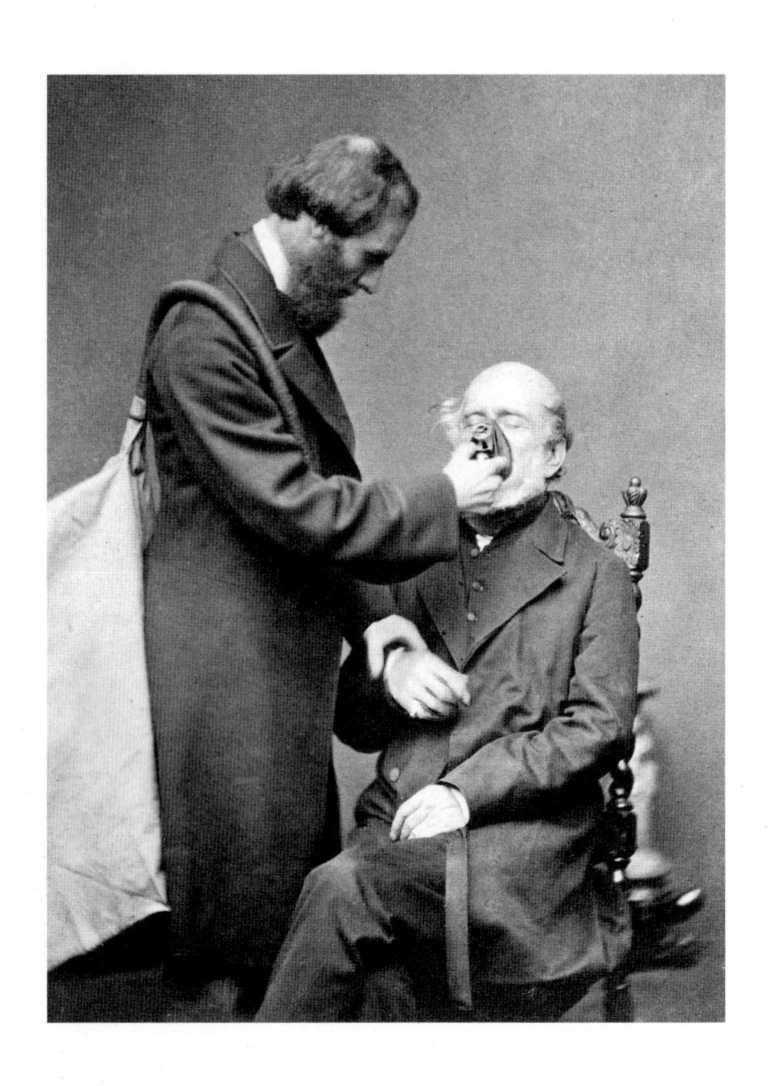

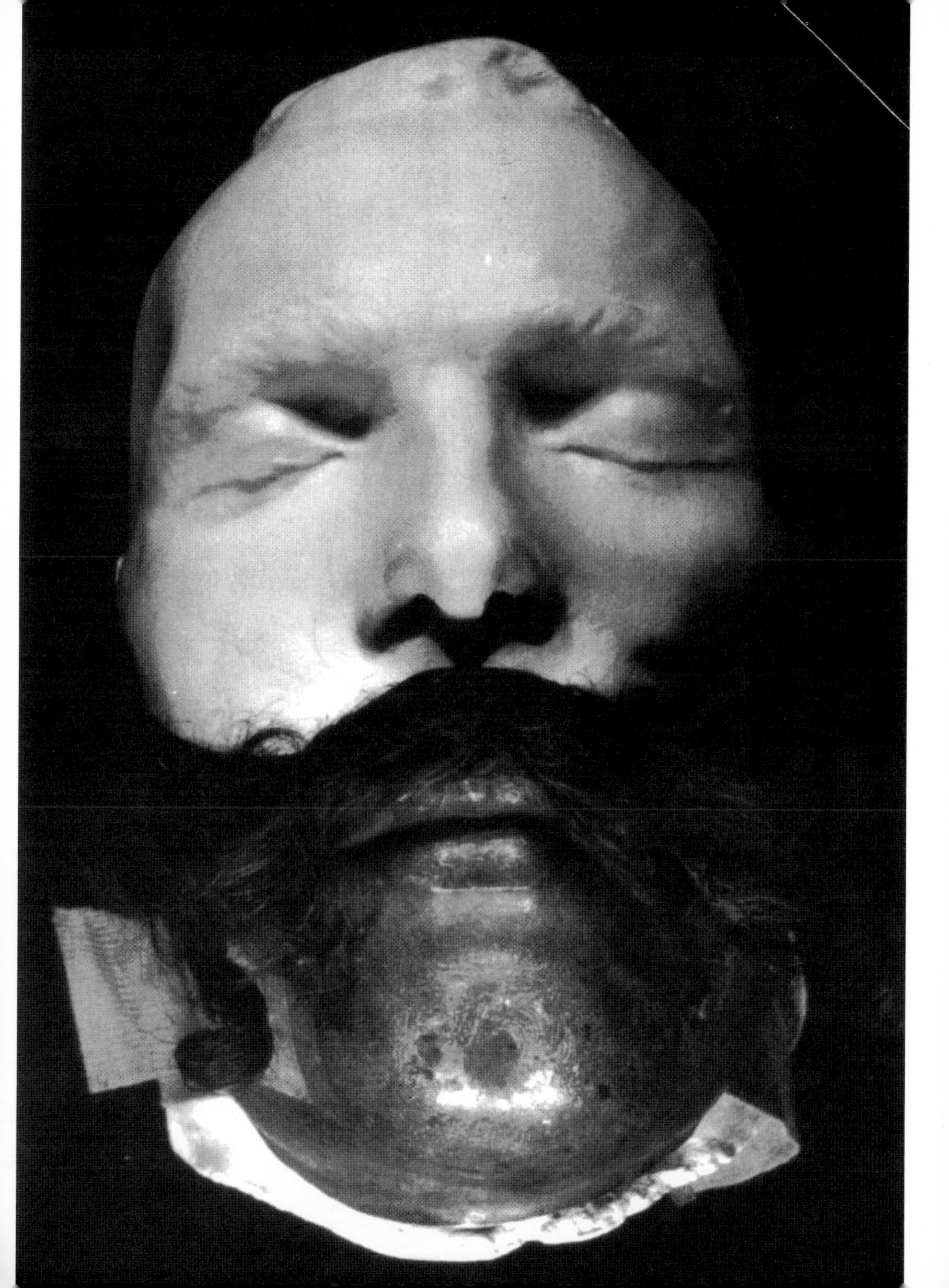

repaired but still appeared disfigured and distorted. Wood, at the Masks for Facial Disfigurements Department (or 'Tin Noses Shop') at the 3rd London Hospital, and Ladd, at the Studio for Portrait Masks associated with the Val-de-Grâce hospital in Paris, were employed in 'aesthetic restoration' for the psychological benefit of the subjects themselves. Veterans were asked to supply a portrait of themselves from before they went to war. These features were then modelled on to a plaster cast of the disfigured face, and a paper-thin mask made from electrotype plate was produced from a mould. The end product was a silvered and painted mask held in place by spectacles soldered to the mask's fragment of nose. These odd sculptures, frail little painted bits of human visages, render identity astonishingly tangible.

Modelling the missing features of the patient as they were before the injury involved all the artistic skills of the portraitist. The system of measurement had to be rigorously applied so that the anatomical volumes of the final mask would fit exactly to the maimed face. The nostrils and mouth were open so as to permit breathing. Painting was no simple matter, either. The masks were first coated in Aspenwall's enamel and finally painted in oil colours treated in a particular way in order to obtain the dull, slightly rough appearance of the skin. As Ladd said:

My objective was not simply to provide a man with a mask to hide his awful mutilation, but to put in the mask part of the man himself – that is, the man he had been before the tragedy.[18]

The Paris studio delivered 67 such prostheses in 1918, and 153 in 1919. The sculptors aimed to achieve the highest possible degree of verisimilitude: 'At a slight distance, so harmonious are both the moulding and the tinting it is impossible to detect the join where the live skin of cheek or nose leaves off and the imitation complexion of the mask begins.'[19] Extraordinarily lifelike as the portrait masks were (even to the point of sewn-in moustaches and special cigarette holders so that the men could smoke as freely as before), many veterans apparently loathed them. Their lack of ventilation made them hot and uncomfortable to wear and their immobility rendered them ultimately mask-like. Andrew Bamji believes there was even a hushed-up riot at Sidcup by those who preferred no face to an abrasive tin equivalent.

During World War II, Harold Gillies' younger cousin, the New Zealander Archibald McIndoe, replaced him as the leading reconstructive surgeon for the Allies; an American physician invented a machine that sliced flesh thinner than tissue paper; and McIndoe founded the celebrated wartime Royal Air Force reconstructive surgery unit at the Queen Victoria Hospital, East Grinstead. There McIndoe and his surgeons reconstructed burns patients by taking cartilage from the hip to form entirely new noses, and skin from the inside of the arm to fashion eyelids. The most famous member of McIndoe's 'Guinea-Pig Club' was Richard Hillary, whose face and hands were badly burned after he was shot down in the North Sea in 1940. After surgery, Hillary toured the UK and the USA and achieved heroic status, and his autobiography, *The Last Enemy* (1947), became a best-seller.[20]

Collaborations between sculptors and surgeons continue today. Artist Paddy Hartley works alongside the biomaterials scientist Dr Ian Thompson at the Oral Maxillofacial Department at Guy's Hospital in London, refining facial implants made of bioactive glass (a material used to repair defects and injuries, which can 'disguise' itself as tissue made by the body) and adapting face corsets into universally fitting facial pressure dressings for the treatment of serious burns. In addition to the medical applications, through the display and wearing of face corsets Hartley examines idealised forms of beauty and alternative means of achieving the wearer's perception of perfection.

Surgeons from Guy's Hospital in London and artists from the Ruskin School of Drawing and Fine Art at Oxford University have noted how individual satisfaction with appearance during the course of cleft palate care has been an important influence upon the psychological well-being of patients, and are researching into whether including artists in a multi-speciality cleft palate team can improve clinical procedures.

mask

Whether worn for the purposes of concealment and/or revelation, masks reveal a tension between impersonality and individuality, conventional representation and likeness. 'Man is least himself when he talks in his own person,' Oscar Wilde wrote: 'Give him a mask and he will tell the truth.' Interest in the mask at the beginning of the last century, and particularly in the 1920s, was noticeable in all the art forms: in fine art, photography, poetry, drama, cinema and dance. One of the most famous of the mask makers of this era was Wladyslaw Theodore Benda, a Polish painter and illustrator who went to New York in 1899. Benda himself described his masks as 'realistic renditions of female beauty... for the most part studies from life,' and later wrote an account of how he made them:

> From careful drawings of the model's face I cut the profile and fit it to the face, filling out the discrepancies with bits of paper until it is all rounded to the contour of the original – in short, a perfect fit. When the whole mask is feature proof, I paint it as would a portrait painter.[21]

Benda's social masks, which were coveted by celebrity groups and high society figures within New York society in the 1920s and 1930s, gave a kind of literal meaning to T. S. Eliot's notion of 'putting on a face to meet the faces...', and were also worn in fashion shots for *Vogue* and *Vanity Fair* during this period. Benda's masks were discovered and

Repairing war's ravages: renovating facial injuries, 1914–18 (pp. 78/79)

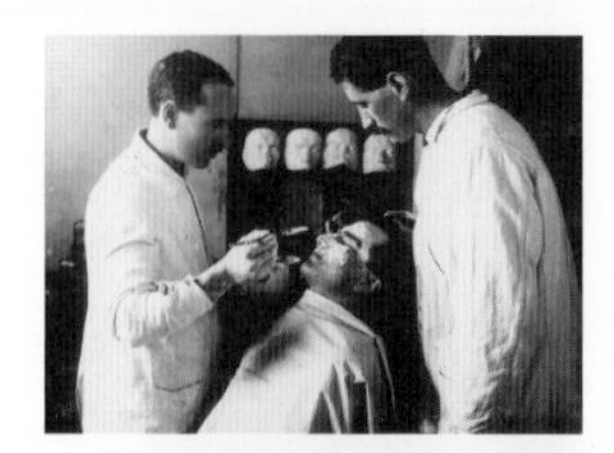

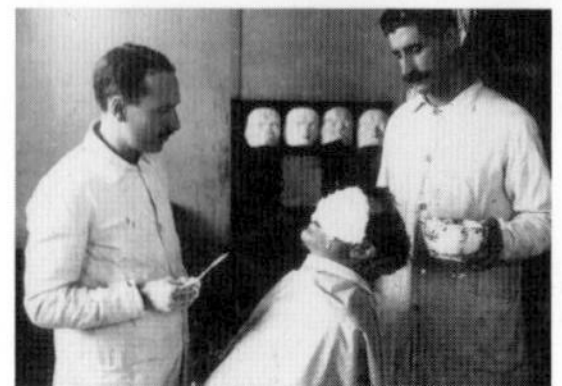

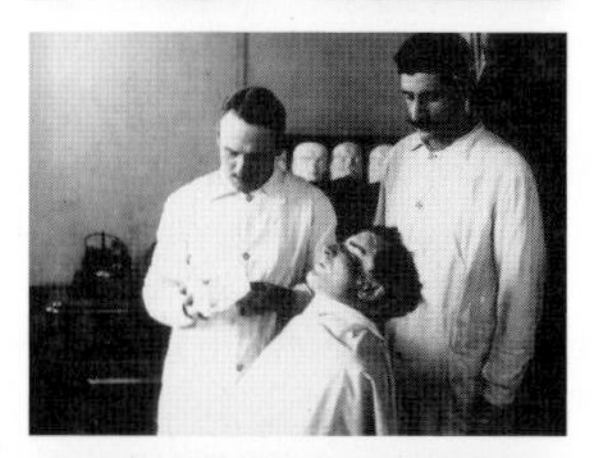

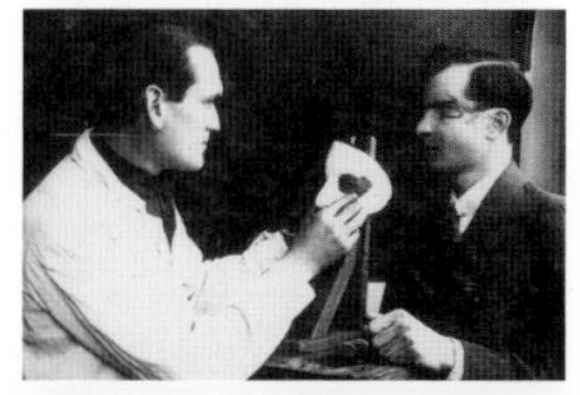

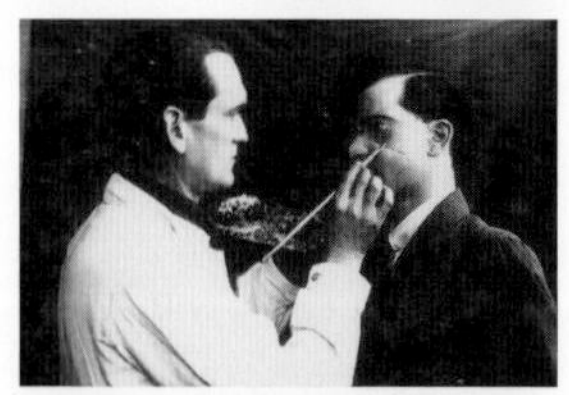

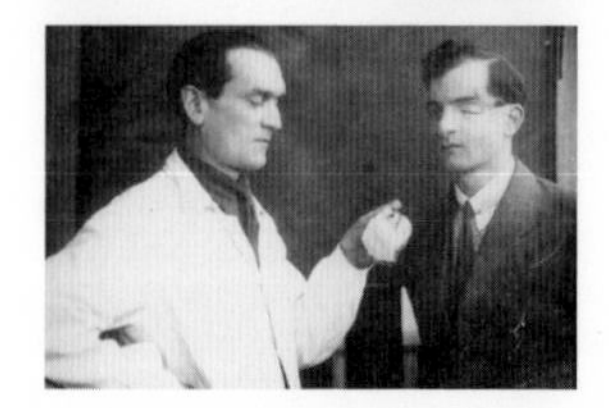

publicised by Frank Crowninshield, the editor of *Vanity Fair*, who later described the first time he saw them:

I was especially enchanted when – on a wind-slit January day in 1920 – (I visited the studio) of Wladyslaw Theodore Benda, for I had heard that he had long pursued the hobby of creating beautiful, varied, and, sometimes, terrifying masks . . . but I was in no way prepared for . . . the breathing beauty that emerged from that gallery of painted faces which – bodiless and without sight – seemed to return my gaze . . . The masks seemed . . . to move; to change with precisely the moods and reactions of living beings. Fresh and faithful renderings of nature (or else bizarre distortions of it), they seemed to breathe and to possess life. In the case of the women's masks, in particular, the skin seemed warm and delicate.[22]

In 1925 the photographer Edward Steichen took an extraordinary series of pictures for *Vogue* in which the models all wore Benda masks. Benda's social masks reside in a category of their own, somewhere between photography, sculpture and cosmetic surgery. (They were used in fashion shots because the texture of the skin, made of laminated paper, appeared more perfect when photographed than that of the skin on the real face.) They are stunningly beautiful, but their popularity was surely dependent on a context in which photography was only just beginning to develop into an artistic medium. It is one thing to wear an African or a theatrical mask, or to make and display a death mask, but what are the cultural implications of keeping an exact replica of your own face in a hat box, and occasionally wearing it to a party? Is this the everyday face as art (a version of Baudelaire's notion of fashion as the distillation of modernity)?

Like Benda's, French artist Jean Pierre Khazem's contemporary, finely sculpted masks of resin or latex are worn by live fashion models in magazines and television commercials. Khazem's masks are an extraordinary blend of the artificial and lifelike. Khazem shows his masks in photographs and videos for galleries as well as for television commercials and magazines. They featured in recent campaigns for Air France and Diesel clothing. His performance work, *Mona Lisa Live*, makes one of the most enigmatic faces in the history of portraiture contemporary through a startling reconstruction. A model wearing a mask of Mona Lisa's face stands on a pedestal beneath a shaft of light in a darkened room. The simultaneously living and simulated effect is characteristic of cultural representation today, especially at the interface of the physical and the computer-generated.[23]

new face

'Tell me what you don't like about yourself' is the surgeon's catch-phrase from *Nip/Tuck*, the US television satire. Sean McNamara, the programme's fictional surgeon, argues:

To succeed these days you need confidence and self-esteem. If you can change a face and change the way someone feels about themselves, that's very satisfying. It's a good thing that we're moving towards a society in which the ways of improving life are available, be it health, life expectancy or, yes, one's looks. Soon the cosmetic surgeon will be regarded as no more than the GP of the aesthetic side of well-being.

The precise origins of plastic surgery (from the Greek *plastos*, meaning framed, moulded, modelled) are unknown. Indian surgeons have been reconstructing noses for centuries following mutilating punishments. As early as 600 BC a Hindu surgeon reconstructed a nose by using a piece of cheek, and by AD 1000, rhinoplasty (nose surgery using skin from the forehead) was known as an Indian technique.[24]

In sixteenth-century Europe, the development of a method of reconstructing the nose by transferring flaps of skin from the upper arm by the Italian Gaspare Tagliacozzi, an early proponent of plastic surgery, may have been due to a rise in facial injuries caused by duelling and street brawling. From the end of the fifteenth century, hereditary syphilis, with its unmistakable depressed or 'saddle' nose, probably created an even more pressing need for treatment. Perhaps the most famous face of syphilis is that of Erik in *The Phantom of the Opera:*

Eyes . . . so deep you can hardly see the fixed pupils. You just see two black holes, as in a dead man's skull. His skin, which is stretched across his bones like a drumhead, is not white, but a nasty yellow . . . and the absence of (a) nose is a horrible thing to look at.[25]

During the nineteenth century, advances in antisepsis and anaesthesia made plastic surgery safer and less painful,and in the 1830s surgeons began refining operations on the sunken saddle nose of congenital syphilis. Popular prejudices about the relationship between the nose, personality, heredity and race pre-dated the pseudo-scientific physiognomic tracts that were to become so popular in the nineteenth century. An 1877 article gives a nonsensical rationale for shaping noses according to types: 'The Roman indicates executiveness or strength; the Greek refinement; the Jewish commercialism or desire for gain; the Snub or Pug, weakness or lack of development.' Such views had been brilliantly satirised in Laurence Sterne's 1759 novel *The Life and Opinions of Tristram Shandy*, where, we read, 'No family, however high, could stand against a succession of short noses.'

In 1887, the New York surgeon, John Roe, perfected a new technique for restructuring the nose without obvious scarring. In the early twentieth century, the German surgeon, Jacques Joseph, became one of the most influential plastic surgeons. His comprehensive handbook of aesthetic surgery of 1931 provided the basic outline for many of the procedures that were to form

Leprosy demon mask (p. 82)
Iron mask, European 1501–1700 (p. 83)

Jean Pierre Khazem, **Mona Lisa live**, 2003 (p. 84)
Paddy Hartley, **Face corset** (p. 85)

the basis of modern plastic or, as it more euphemistically came to be known, 'aesthetic' surgery.

Surgical correction of this kind resulted in improvements to confidence and self-esteem. It also became something of a public spectacle in the manner of the earliest anatomical dissections, literally performed in theatres (from which the modern term 'operating theatre' derives). In 1931, Dr John Howard Crum performed the first public face-lift in the Grand Ballroom of the Pennsylvania Hotel in New York. Members of the audience intermittently fainted and a pianist played jaunty popular tunes. But, as critic Erin O'Connor points out in a fascinating online article on modern plastic surgery, 'The Love Song of Plastic Surgery', what is significant in the preceding developments of the nineteenth century is the justification of plastic surgery on the grounds of its psychiatric benefits: that it supposedly repaired the psychic distortions produced by the ill-formed feature. In making this claim, plastic surgery placed itself in direct competition with the emerging practice of psychotherapy.

From the first, plastic surgeons regarded themselves as sculptors of living flesh, breaking new aesthetic ground – hence the increasing use of the word 'aesthetic' instead of 'plastic'. The metaphor of the surgeon as sculptor was first used by Roe, but it quickly came to dominate popular and medical writing about plastic surgery. In 1951, the surgeon Murray Berger explained that:

The well-informed plastic surgeon – essentially a sculptor – studies the proportions of his patient's facial features . . . He has a sculptor's concept of the features that form the particular face under study and consequently can obtain a gratifying surgical result.[26]

During this period surgeons emphasised their artistic and aesthetic skills as much as their medical expertise. As cultural historians have noted, the idiom of high art was useful to help plastic surgery claim its place as an elite branch of medicine. It was also necessary for surgeons to establish confidence in their artistic skills, because alongside its healing role, plastic surgery had commercial potential.[27]

Bizarre as it may seem to us now, in the 1920s and 1930s plastic surgeons like the American Vilray Papin Blair published images of classical statuary alongside photographs of their patients before and after surgery, citing the attainment of the 'classic Greek' face as the goal of surgery. For Blair, this was 'the most beautiful type ever recorded'. In the following decades, New York surgeons like Polish-born Jacques Maliniak and Maxwell Maltz also used the classical statue as a touchstone. Maliniak published a photograph of the Venus of Capua in his *Sculpture in the Living: Rebuilding the Face and Form by Plastic Surgery* (1934). Maxwell Maltz's *New Faces – New Futures* (1936) features 'before and after' photographs of a portrait bust of Cleopatra subjected to imaginary surgery to correct her long hooked nose. Maltz also prints a photograph of the fourth-century BC Greek sculptor Praxiteles' Aphrodite of Cnidos to exemplify the classic Greek nose.

By 1930, American *Vogue* recognised Venus as the model Hollywood face:

Nowhere else in the world are there gathered together so many conventionally beautiful people. This is a town inhabited almost entirely by gods and goddesses of beauty. The girl shutting the window is Venus disguised as a most exquisite Madonna. The newspaper boy is a fair young Apollo . . . After a time, one loses one's sense of proportion and nothing remains to stare at.[28]

The development of aesthetic surgery pioneered in those early operations on World War I veterans required not only the discovery of anaesthesia, but also a shared social view of beauty and of what was considered a 'normal' or 'acceptable' appearance. From the eighteenth century onwards, so-called 'Enlightenment ideas', supported by the impact of nineteenth-century physiognomics, equated beauty with virtue and well-being. The more sinister corollary of this was that its opposite was also believed to be true. As is shown by *Making the Body Beautiful*, a recent cultural history of cosmetic surgery by the critic Sander Gilman, prejudice over racial differences increasingly focused on appearance. In the late nineteenth century, racial prejudice against the Jew was increasingly based on a 'science' of race which purported to find evidence of physical inferiority in the dispensation of the face (for example, the Jewish nose was supposedly an example of this kind of physical classification). In this way, 'self' image is never isolated; it is always derived from a collective or shared ideal and a cultural context. Plastic surgeons in the United States and Europe were uniquely positioned to mask the stigma of socially unacceptable diseases such as syphilis, and were even seen as the potential saviours of criminals and the poor. They were also able to 'help' Jews, light-skinned blacks and pug-nosed Irish 'pass' as 'white' and avoid the ethnic and racial hatred triggered by their non-Caucasian faces. However, aesthetic surgery of this kind, carried out to mask racial features and types, begs the question of whether the supposedly post-operative improvement in self-confidence and esteem is in fact outweighed by the psychological burden of the suppression of social identity.

According to Gilman, Britons now spend more than £225 million a year on cosmetic surgery. It is one of the most common reasons given by women for non-property loans, and one in twelve cosmetic surgery patients are men. In 2002 alone, some 25,000 cosmetic surgery procedures were conducted in the UK, and 860,000 in the USA. A list of common procedures today would include: cheek and chin implants (for the augmentation of the face); collagen and fat injections (which

Helena Rubinstein, **Pomade noir masque**, 1939 (p. 88)
Helena Rubinstein, **Masque treatment**, 1939 (p. 89)

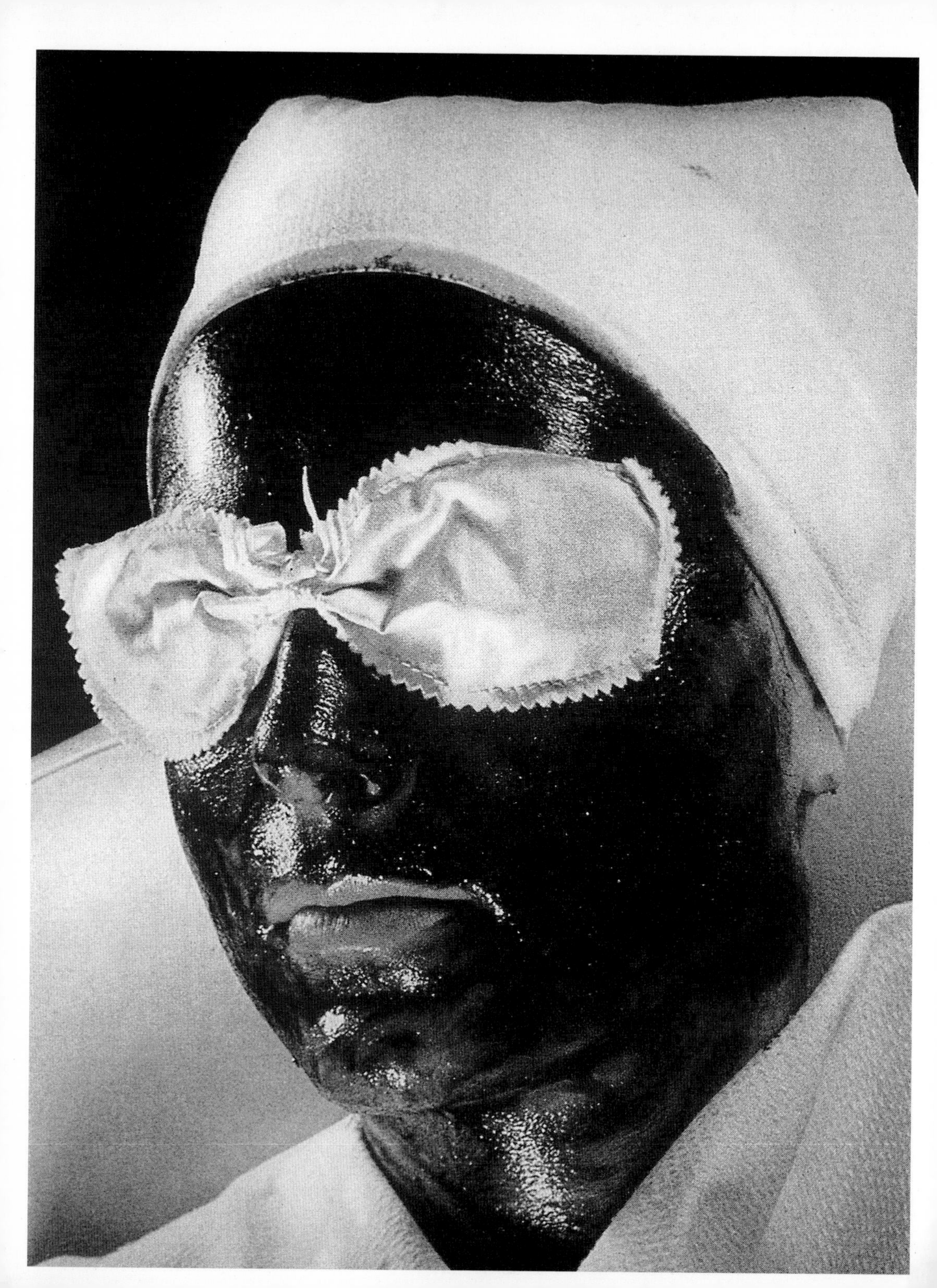

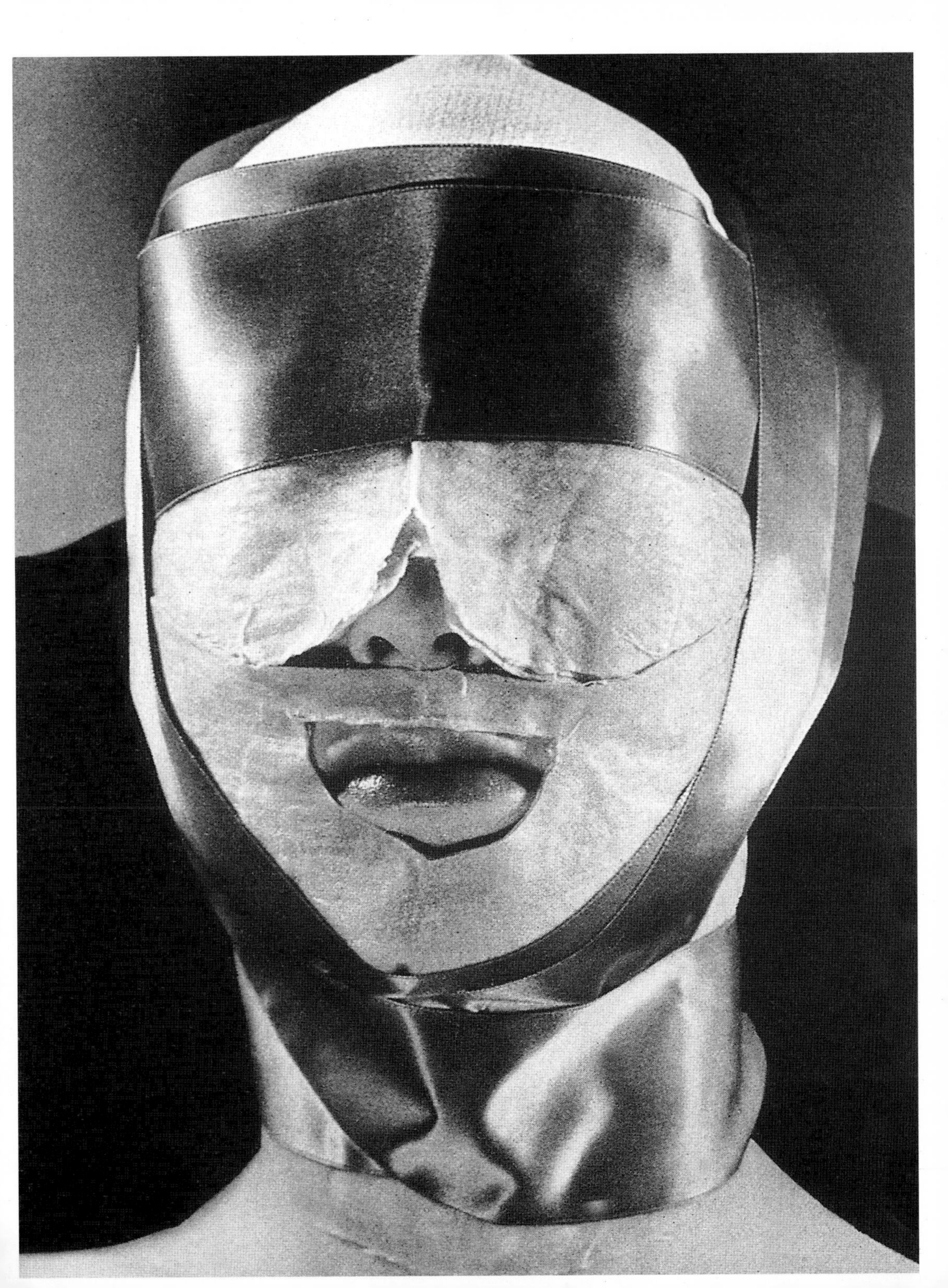

enhance the lips or plump up sunken facial features); ear pin-back, or otoplasty (which brings the ears closer to the head); eyelid tightening, or blepharaplasty (which eliminates drooping upper eyelids and puffy bags below); face-lift, or rhytidectomy (which tightens the jowls and neck); forehead-lift (which minimises creases in the forehead and hooding over the eyes); hair transplantation; nose surgery, or rhinoplasty (which changes the appearance of the nose); scar revision and the removal of birthmarks and skin resurfacing by laser; chemical face peel and dermabrasion, or sanding of the skin (which removes wrinkles, skin blemishes, etc.)[29] With the current trend for copying celebrities, in 1998 two well-known American cosmetic surgeons, Dr Richard Fleming and Dr Toby Mayer, started compiling a list of stars' body parts most requested by patients. These included:

Most requested features (women)

Nose:	Nicole Kidman, Reese Witherspoon, Diane Lane
Hair:	Jennifer Aniston, Debra Messing, Sarah Jessica Parker
Eyes:	Halle Berry, Jennifer Lopez, Cameron Diaz
Lips:	Liv Tyler, Uma Thurman, Renée Zellweger
Jawline/chin:	Salma Hayek, Julianne Moore, Kim Cattrall
Cheeks:	Jennifer Lopez, Halle Berry, Jennifer Garner
Sculpting:	Angelina Jolie, Britney Spears, Jennifer Lopez
Skin:	Michelle Pfeiffer, Gwyneth Paltrow, Sandra Bullock

Most requested features (men)

Nose:	Ben Affleck, Edward Burns, Jude Law
Hair:	Pierce Brosnan, Richard Gere, Hugh Grant
Jawline:	Johnny Depp, Matthew McConaughey
Lips:	Brad Pitt, Ralph Fiennes
Cheeks:	Leonardo DiCaprio, George Clooney, John Corbett
Chin:	Russell Crowe, Kiefer Sutherland, Matt LeBlanc
Sculpting:	Tom Cruise, Benjamin Bratt, Matt Damon
Skin:	Ethan Hawke, Hayden Christensen, Ryan Phillippe

George Orwell's adage that by the age of 50 we have the face we deserve is now, perhaps,outdated. These days, we can have almost any face we would like, whenever we want, provided that we can afford it. It is fascinating that those who undergo emulation surgery are attempting to copy screen faces which have themselves already been augmented at a number of different levels (for example, make-up, styling, lighting and – sometimes – extensive digital or cosmetic enhancement). Some people have even chosen to have their faces remodelled to look like their favourite animals, in a bizarre reversal of Darwinian notions of the evolution of the species. Whatever the aesthetic considerations, in an interview with the *Independent*, twin plastic surgeons Maurizio and Roberto Viel from the London Centre for Aesthetic Surgery describe how they still see the science of plastic surgery inextricably intertwined with art. Roberto says:

I like to think of this as the artistic field of medicine, in which all the medical training is necessary, but in which you also need a certain sense of beauty. Maurizio and I trained in medicine, but our mother is an opera singer, so we grew up with an artistic leaning which we want to express.

One of the most recent forms of plastic surgery is Isologen, a therapy that uses parts of the body to cure or heal other parts. The procedure involves taking a tiny sample of skin (usually from behind the ear) and using it to culture more skin cells, which are then injected into the creases or scars that need attention. By injecting cultured cells into the wrinkles and depressions in the face, new collagen is produced, which in turn gives the skin renewed structure, pushing out and reducing wrinkles and scars. As tissue transplants, micro-surgical skills and anatomical knowledge of the face develop, face transplants move out of the territory of science fiction to become an ever-increasing surgical possibility.

someone else's face

A face transplant would involve removing the face, facial muscles and subcutaneous fat from the recipient. A line would be traced across the hairline and down the temples on both left and right sides. It would pass just in front of the ears and round the jawline. The donor face from a recently dead person, complete with lips, chin, ears, nose, eight major blood vessels and even some bone would then be grafted into place. Human face transplants have not yet been legalised in Britain, although a team of surgeons is poised to perform the first face transplant in America. And a number of animal face transplants have already been successfully carried out around the world.[30]

There have, however, been a number of groundbreaking operations in which a person's own face has been successfully reattached or 'replanted'. In 1994, nine-year-old Sandeep Kaur got her hair caught in a piece of farm machinery. As a result, her entire face and scalp were ripped off. Both were successfully replanted by microsurgeon Abraham Thomas. And in 1998 the face and scalp of a 28-year-old Australian woman were torn off in another accident involving farm machinery. Though she was left with only her chin, left ear and lower lip intact, the rest of the woman's facial features and scalp were recovered from the machinery, packed in ice, and subsequently reattached to her skull.

When asked to what extent the woman's facial appearance would change as a result of the surgery, Professor Wayne Morrison, who led the team of surgeons at St Vincent's Hospital, Melbourne, in the 25-hour operation, pointed out that much of human physiognomy is based on what takes place beneath. 'The facial expression and the movement of the face is more to do with what is underneath the skin, and the patient on to whom you transplant a face would retain that sort of movement.' He went on to say that the prospect of taking a dead person's face and draping it over the skull of a living man or woman was now much closer: 'It is simply like changing the cloth of an old armchair.' But the *issues* involved in 'draping [a dead person's face] over the skull of a living man or woman' are of course more complex. Professor Morrison also described the operation as the most horrific thing he had ever seen. Transplanting a dead face removed under pristine surgical conditions is arguably easier than replanting one that has been ripped off and possibly crushed, and that needs to be trimmed and cleaned before the operation. A recent report published by the Royal College of Surgeons in London, however, concluded that more research is needed on the psychological impact both on the recipient of a face transplant and on the donor's family before carrying out face transplants in Britain. Physical problems associated with possible organ rejection and immunosuppressant drugs are a particular cause for concern owing to the very large areas of skin involved – as they are to a lesser extent with other transplants.

Does the shiver down the spine or the heave of nausea one feels on reading about the terrible accidents above derive in part from the irrational sense that without their faces the victims are somehow no longer human, have no identity? That a face is congruent with an inner self or soul ('what is underneath the skin'), and that external and internal selves are joined together by the face that confers meaning on them? In an interview with David Concar for *New Scientist* in May 2004, even Serge Martinez, one of the team of North American scientists who have been practising for face transplant operations by swapping the faces of bodies donated for medical research, expresses a sense of mystery. 'It is really awesome to lift up a whole face and lay it back down,' he says. Jenny Diski's fascination with Bella's face 'with all the life subtracted', quoted at the start of this chapter, has been shared by generations of artists, anatomists, forensic experts and surgeons who have probed beneath the surface of the face and mapped its contours both inside and out, on the living and on the dead. For all of them, the fascination and mystery of the face exists in the combination of interior and exterior, consciousness and countenance. 'As far as I'm concerned I firmly believe that man's soul is housed in his skin ... It's no metaphor,' the molecular chemist 'K' tells the narrator in Kobo Abe's novel *The Face of Another*. Or, as the contemporary philosopher and medical doctor François Dagonet argued in 1992, 'The psyche emerges from the most complex corporeal structures.'

Nōh mask of a young woman (p. 94)

Leiter of Vienna, **Bakelite face phantom for practising eye operations**, circa 1860–90 (p. 95)

Corinne Day, **Me just before brain surgery, London Hospital 1996**, 2000 (p. 96)

Max Factor beauty calibrator, 1932 (p. 97)

The portrait . . . was that of a young girl. It was a mere head and shoulders, done in what is technically termed a vignette manner; much in the style of the favourite heads of Sully . . . The frame was oval, richly gilded and filigreed in Moresque. As a thing of art nothing could be more admirable than the painting itself. But it could have been neither the execution of the work, nor the immortal beauty of the countenance, which had so suddenly and so vehemently moved me. Least of all, could it have been that my fancy, shaken from its half slumber, had mistaken the head for that of a living person. I saw at once that the peculiarities of the design of the vignetting, and of the frame, must have instantly dispelled such an idea – must have prevented even its momentary entertainment. Thinking earnestly upon these points, I remained, for an hour perhaps, half sitting, half reclining, with my vision riveted upon the portrait. At length, satisfied with the true secret of its effect, I fell back within the bed. I had found the spell of the picture in an absolute life-likeness of expression, which at first startling, finally confounded, subdued and appalled me.

Edgar Allan Poe, 'The Oval Portrait'[1]

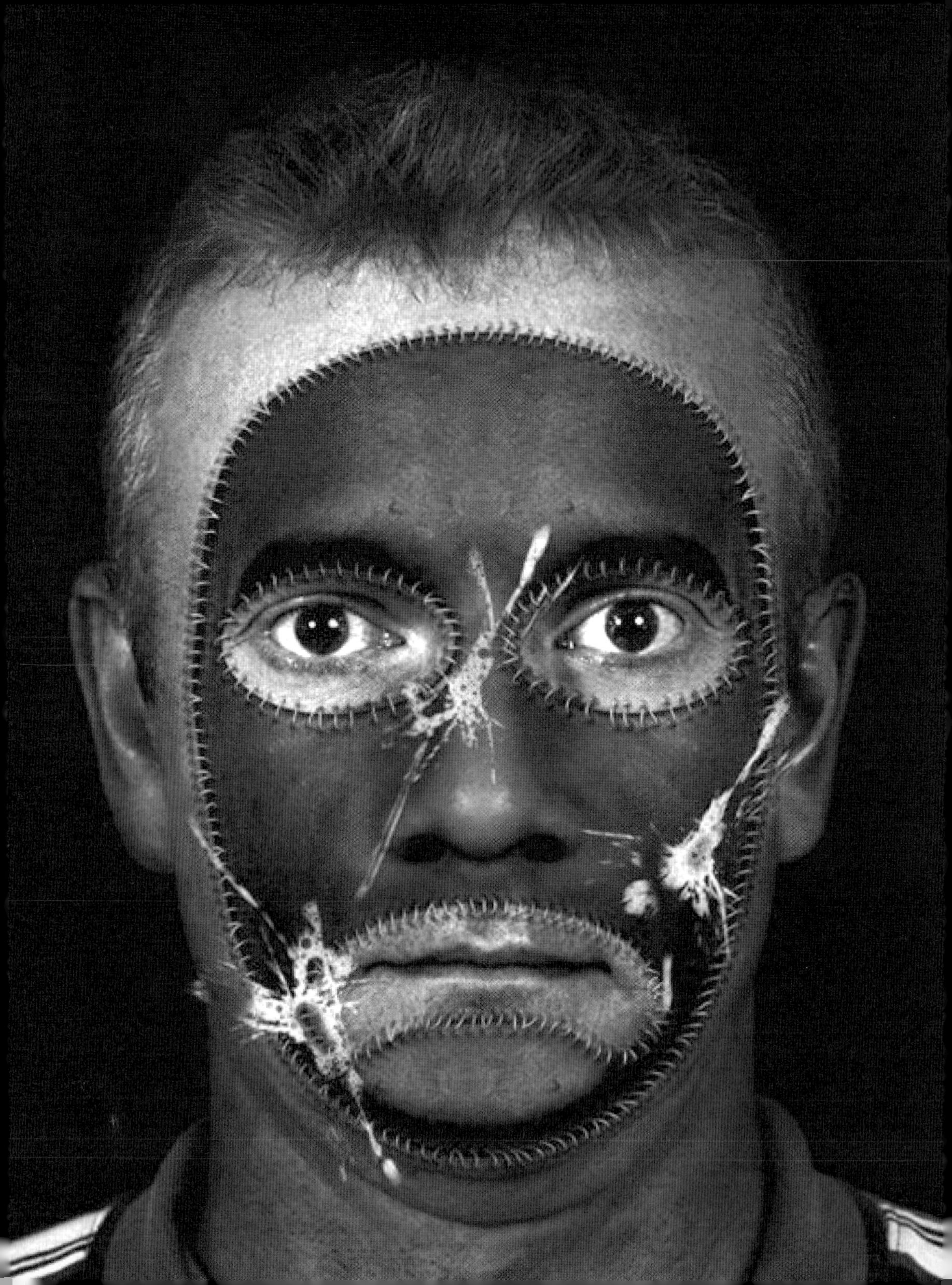

data face

'Classification' theories and science and technology have always played a significant role in characterising the face. From Ancient Greece to our modern day, the classification of the beautiful (and 'good') face has been based on balance and symmetry. According to the evolutionary psychologist Steven Gangestad, 'Symmetry alone explains why Elizabeth Taylor, Denzel Washington and Queen Nefertiti are universally recognised as beautiful.'

Matthew Cornford and David Cross's *10* was staged in Derby in 1998. Cornford and Cross had put up posters advertising a 'unique beauty contest ... looking for the face of Derby pride'. Almost 200 people turned up to be photographed (straight on, in harsh light), and then have their image judged by a computer. Ten winners were selected and put in order according to the 'percentage of their beauty'. The finished artwork consists of the winning mug-shots blown up to billboard size. 'When we had the final selection of faces, they seemed bland,' said Cross. 'Some of the women, in particular, looked so similar as to be easily confused.'

Harwood, **White man, black mask**, 1996

Why is this? Scientific research suggests that symmetry – the classical requirement of beauty advocated by everyone from Plato onwards – remains the one aesthetic constant, so Cornford and Cross had programmed their computer to pick out the most symmetrical faces. They used technology developed for digital 'fingerprinting' and also gave the computer a control: a kind of androgynous über-beauty created by melding a range of famously beautiful faces, from Nefertiti to Keanu Reeves. The computer took 24 separate facial measurements from each contestant, compared their faces to the 'ideal' of symmetry and proportion, and then selected the winner from its own numerical ranking of the scores.

There is something almost sinister about the chosen images: their uniformity is reminiscent of those Nazi photographs of pristine Aryan specimens whose facsimile features make them appear like androids. In contrast, when the Nazis photographed the 'rejects' in the concentration camps, the photographs were full of life in all their individual and quirky diversity – or in other words, their lack of symmetry.

Recent research suggests that we are genetically programmed to prefer symmetrical faces. And certainly, although there are wide differences between individuals and across cultures in what is considered physically attractive, there is a remarkable consensus of agreement about what constitutes facial beauty. For the Ancient Greeks, facial beauty was a question of proportion. In the mid to late fifth century BC the sculptor Polyclitus's influential book entitled *Canon* offered a mathematical and proportional system for sculpting beauty. He used symmetry to reflect a highly complex system of measurements and internal proportions. The book and its details are lost, but Polyclitus's sculpture Doryphoros embodied his theories. Medieval artists believed that the perfect face was neatly divisible into sevenths. As late as the eighteenth century, Sir Joshua Reynolds still believed that beauty was simply a matter of physical proportions. Even Hogarth believed he had found the overriding principle in 'the wavy line of beauty': 'the greatest, indeed the indispensable element of all beautiful things is the smooth serpentine line'.

In 1990, as part of the congress 'Art and Life in the 90s' in Newcastle, the performance artist Orlan began a project entitled *The Reincarnation of St Orlan*. This involved undergoing a sequence of operations to remodel her own face according to a synthesis of iconic female prototypes drawn from art and mythology, as a parody of the way male artists have fragmented and fetishised the female body in their work. Each operation was designed to recreate an individual facial feature, and was filmed in graphic detail. Orlan herself choreographed each operation, which was conducted by surgeons working from a computer-generated image created by herself. This composite image used the nose of a sculpture by an anonymous School of

Fontainebleau artist, the chin of Botticelli's *Venus*, the forehead of Leonardo da Vinci's *Mona Lisa*, the eyes of François Gérard's *Psyche* and the mouth of Gustave Moreau's *Europa*. When Orlan's face was prepared for one of these operations, the markings were reminiscent of the illustrations in medieval and physiognomic texts.

Since 1998, Orlan has worked on a series of digitally manipulated self-portraits, *Self-Hybridations*, which use morphing software to produce virtual versions of her own face (instead of actual surgery), incorporating concepts of body art and transformation from pre-Columbian Mayan and Olmec cultures.

'has baby a clever head?'

Indeed, for Orlan, as for many, the way we look is dictated by socio-cultural as well as genetic influences. Advances in the methods that can be used to determine the constituent parts of facial identity must await further research into the best way to measure and compare faces. It is also the case that the stereotypes and prejudices associated with facial appearance (the 'criminal' face, the 'intellectual' face, and so on) have arisen from nineteenth-century scientific theories that are now regarded as at best debatable (not least because of their implicit association with the discredited theory of eugenics).

The first of these theories is, of course, physiognomy, said to have begun with Pythagoras, whose views are developed in *Physiognomica* (sometimes attributed to Aristotle). The term is derived from the Greek *phusis* (nature) and *gnomon* (a judge or interpreter). Medieval treatises divined traits of character from both human and animal physiognomies, relating features and facial types to astrological signs and characters. Thus, full lips supposedly revealed sensuality, while a wide-eyed or owl-like person was clever. Someone who resembled a fox was deceitful, and a lion, strong. Physiognomy sought to locate what Leonardo da Vinci had called the 'relevant traits' of the human face and to map character or personality through a classification of facial features and their corresponding moral traits.

The notion that a person's true character is revealed in their face has remained a striking concept for the Western imagination ever since. The basic ideas of physiognomics connect to 'humoral' theory – the notion that four basic physiological 'humours', or essential states, condition human behaviour. The stock facial expressions of classical Greek, ancient Chinese, Japanese Nōh and Indian Kathakali theatre are influenced by physiognomic theories. According to this view, the face is not unlike a mask, which may be used to both hide and reveal a person's inner self, depending how it is 'read'.

In the nineteenth century, variations on the theory of physiognomy dominated

Orlan having cosmetic surgery in New York, 1993 (p. 104)
Wax vanitas life/death head, circa 1700–1800 (p. 105)

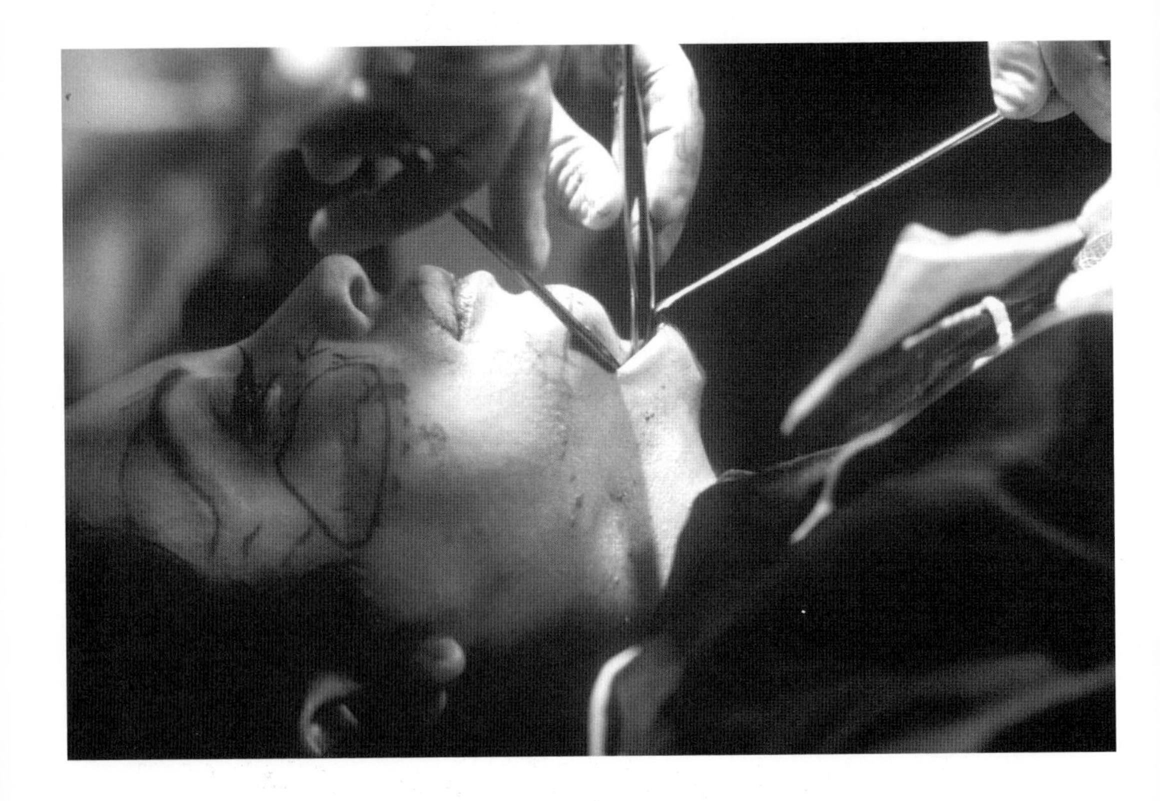

Vanitas
Vanitatum & oña
Vanitas

culture and literature. The four-volume *Physiognomische Fragmente, zur Beförderung der Menschenkenntnis und Menschenliebe* (1775–8) ('Essays on Physiognomy: Designed to promote the Knowledge and Love of Mankind') by the theory's main proponent, Johann Kaspar Lavater, became the most influential text. Lavater, a Swiss theologian, physiognomist and poet, believed that there was a correspondence between physical appearance and moral character, 'between the external and internal man, the visible superficies and the invisible contents'. Facial beauty was a sign of virtue, and ugliness was evidence of vice. 'The moral life of man,' he wrote, 'reveals itself in the lines, marks and transitions of the countenance.' Lavater's ideas influenced conceptions of appearance and character for over a century and are still embedded in facial stereotypes today.

The other nineteenth-century 'face' sciences were phrenology (the 'science' of looking at the skull as a source of character revelation), Darwin's evolutionary theories, and the criminal anthropology of the nineteenth-century Italian criminologist Cesare Lombroso. All three sciences used their subjects as proof of what they already expected, or as raw material by which to promote their particular agendas. Each extrapolates from observation of human difference to deduce rigidly limited theories in a manner that palaeontologist and evolutionary biologist Stephen Jay Gould describes as 'The tenacity of unconscious bias and the surprising malleability of quantitative data in the interests of a preconceived idea.'[2] In *Criminal Man* (1876), Lombroso derives his evolutionary theory of hereditary criminality from looking at a skull:

At the sight of that skull (of the brigand Vihella), I seemed to see all of a sudden, lighted up as a vast plain under a flaming sky, the problem of the nature of the criminal – an atavistic being who reproduces in his person the ferocious instincts of primitive humanity and the inferior animals.[3]

Examining and collecting skulls was a hobby throughout the Victorian period, and 'phrenology' was still fashionable in the early twentieth century. One eminent American advocate of phrenology in the nineteenth century, Orson Squire Fowler, believed that a phrenologist could 'pronounce decisively whether a man is a liar, a thief or a murderer' through the measurement of his skull, while in its June 1901 issue, *Strand Magazine* published an article which (in all seriousness) attempted to 'apply our knowledge to the solving of that all-important question, "Has Baby a clever head?"'

Lombroso postulated a meaningful similarity between the facial asymmetry of criminals and flatfish with both eyes on the upper surfaces of their bodies. Darwin, too, sought to understand the semiotics of facial expression by comparison with animals; on one occasion he visited a zoo and put a freshwater turtle in with the monkeys to see if they expressed surprise. He wanted to discover why certain movements communicate expressions. Half a century later, the art historian Ernest Gombrich was still considering human facial expression by comparison with animals:

Unless introspection deceives me, I believe that when I visit a zoo my muscular response changes as I move from the hippopotamus house to the cage of the weasels . . . this doctrine relies on the traces of muscular response in our reaction to forms; it is not only the perception of music which makes us dance inwardly, but the perception of forms.[4]

All these theories are heavily dependent on Lavater's physiognomy: the belief that character was indicated by facial features. In the mid-nineteenth century, physiognomy was offered as the new science of diagnosis, like the stethoscope, or chemical analysis in medicine, as part of a true 'science of character'. The scientific age provided information and methodologies that artists looked to for assistance in their portrayal of nature.

The history of the interplay of art and science in theories of the face is crucial. Throughout the late nineteenth century, criminal anthropology, psychiatry, art and physiognomy were aligned by their attempts to interpret and represent the face. As a result of the currency of theories of the classification of the face and the new face sciences, the Victorians were preoccupied with faces and were also expert face readers. In the nineteenth and early twentieth centuries the average person was more inclined than they are today to apply moral judgement to the face and to attribute personality types to particular configurations of facial features. Every feature had a specific meaning. Although there are lingering stereotypes, we no longer scrutinise and record every detail and deduce character in the same way.

classified face

Transcribing and recording the face also played a key role in developing nineteenth-century thinking about character and its classification. Social portraiture and scientific photography represent different ways of exploring the fabric of society. Recent books by the critics and curators Robert Sobieszek, Roger Hargreaves and Peter Hamilton have demonstrated how conventions established in one area influenced modes of depiction in another. Fascination with phrenology and physiognomy influenced styles in studio portraiture as well as characterising the frames of reference for and depictions of other cultures, mental illness and criminals.

Dr Hugh Diamond, Superintendent at the Surrey County Asylum in the 1850s, used photography to illustrate and record mental illness. In a paper delivered to the Royal Society in 1856, entitled 'On the Application of Photography to the Physiognomic and Mental Phenomena of Insanity', he outlined his belief that research into mental illness

G. E. Madeley, after G. Spratt, **Physiognomist**, 1831 (p. 108)

T. Rowlandson, **Franz Joseph Gall leading a discussion on phrenology**, 1808 (p. 109)

THE PHYSIOGNOMIST.
G Spratt del.
G.E. Madeley, lith. 3 Wellington St. Strand.
Published by C Tilt, Fleet Street. 1831.

AWERS. THIEVES & MURDERERS.
OETS. DRAMATISTS. ACTORS.
PHILOSOPHERS.
STATESMEN & HISTORIANS

could be advanced through the photographic transcription of the facial expressions of his patients:

An Asylum on a large scale supplies instances of delirium with raving fury and spitefulness, or delirium accompanied with an appearance of gaiety and pleasure in some cases, and with constant dejection and despondency in others, or imbecility of all the faculties with a stupid look of general weakness, and the Photographer enables in a moment the permanent cloud, or the passing storm or sunshine of the soul and thus teaches the Metaphysician to witness and trace out the visible and invisible in one important branch of his researches into the Philosophy of the human mind.

The carte de visite portraits of the patients in the West Riding Pauper Lunatic Asylum at Wakefield produced by its director, James Crichton Brown, or the commercial photographer Henry Herring's 'before and after' portraits of mentally ill patients at the Bethlem Royal Hospital, Beckenham, Kent also reveal the connection between photography's artistic uses and its role in providing medical and scientific evidence.

As the nineteenth century progressed, there was also an ever-widening deployment of both the popular conventions of physiognomy and portraiture and the technologies of visualisation as tools of surveillance and classification. Taxonomy (the science and practice of classification) underpinned modes of evolutionary thinking, social analysis and early modern thought. Investigation into the physical and cultural development of Victorian Britain was structured by the presentation of difference from a perceived norm. Photographic portraits doubled as both documentary evidence and analytical data. William Henry Fox Talbot, the nineteenth-century English physicist and pioneer of photography, described the photograph as 'the pencil of nature' and asserted its usefulness for the 'inductive methods of modern science'. The term 'photography' soon became a synonym for objective knowledge, and belief in its documentary powers remains part of the photographic stereotype today. Portraits were used by the police, doctors and anthropologists alike to create visual archives for classification and control.

Encyclopedic publications of anthropological photographs typify this trend in the late nineteenth century: for example, the eight volumes of *The People of India: A Series of Photographic Investigations* (1868–75), edited by Dr John Forbes and Sir John William Kaye; John Thomson's *Illustrations of China and Its People* (1873–4), and Edward S. Curtis's *The North American Indian, Being a Series of Volumes Picturing and Describing the Indians of the United States and Alaska* (begun in 1896, published 1907–30). Through this kind of measurement, classification and cataloguing of indigenous peoples throughout the world, ethnographic photographers, influenced by

Cesare Lombroso, **Portraits of German criminals**, circa 1895 (p. 112)

Known militant suffragettes' police ID cards, 1914 (p. 113)

physiognomic and phrenological theories, could be said to have created the prototype of what we now know as the classic 'mug-shot'.

According to records in Europe, from the 1840s onwards photography was also used in order to record and disseminate likenesses of criminals. The Birmingham police force was one of the first British forces to do this. From around 1870, a photograph of every boy who was taken in by Dr Barnardo's Home for Destitute Lads in England was filed with a brief note of his history up to that point. This information was used by the authorities to help find boys who ran away and to keep a judicial record.

But there was no standard technique. Criminal portraits, or 'mug-shots' as they came to be known in America, were no different from other kinds of portraits and were often highly stylised, with background props (sometimes resembling a theatrical set) and exaggerated expressions. The later mug-shot convention of full-face and profile had not yet been established. From around 1860, however, the use of head restraints and other restraining devices suggests that attempts to standardise the judicial image were being made.

In 1856 the French journalist Ernest Lacan advocated that the system of keeping photographic records of criminals should be expanded:

Which outlaw could thus escape the vigilance of the police?... his portrait would be in the hands of the authorities; he would not be able to escape and would be forced to recognise himself in this accusing image.

In 1883 Alphonse Bertillon, a physiognomic anthropologist working in France, developed the photographic identification system which was used by many international police forces for nearly a century. Photographs of the head, taken in profile and full face, were set against a neutral backdrop and printed on a card which listed other pertinent details for the purposes of identification (for example, date of birth, height, hair colour and so on). These cards were sub-divided into three groups based on head length (large, medium and small) and filed accordingly.

Bertillon also published a 'Synoptic Table of Physiognomic Traits', which contained hundreds of photographs, broken down into tiny details of the features of the male human face, as these were the 'marks' by which the criminal would be identified. In particular, Bertillon believed that the ear provided a wholly reliable measure:

It is nearly impossible to encounter two ears that are identical in each of their parts, and the numerous formal variations that make up that organ seem to continue without modification from birth to death.[5]

Analysis of the shapes of ears by Sherlock Holmes appears in Sir Arthur Conan Doyle's

Francis Galton, **Family composite portraits**, from Galton's *Inquiries into the Human Faculty and Its Development*, 1883 (p. 114)

Alphonse Bertillon, **Type portraits**, from *La Photographie Judiciaire*, circa 1890 (p. 115)

Stab. V. TURATI inc.

1
2
3
4
5
6
7
8
9
10

SPECIMENS OF COMPOSITE PORTRAITURE

PERSONAL AND FAMILY.

Alexander the Great From 6 Different Medals.

Two Sisters.

From 6 Members of same Family Male & Female.

HEALTH.

23 Cases. Royal Engineers. 12 Officers. 11 Privates

DISEASE.

6 Cases

9 Cases

Tubercular Disease

CRIMINALITY.

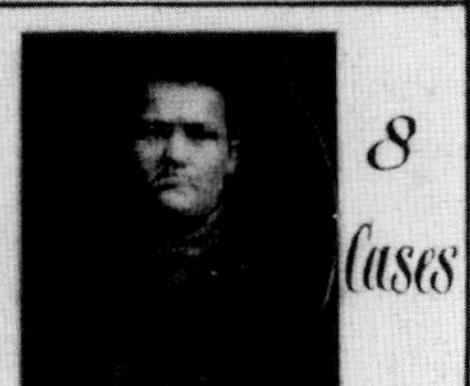

8 Cases

4 Cases

2 Of the many Criminal Types

CONSUMPTION AND OTHER MALADIES

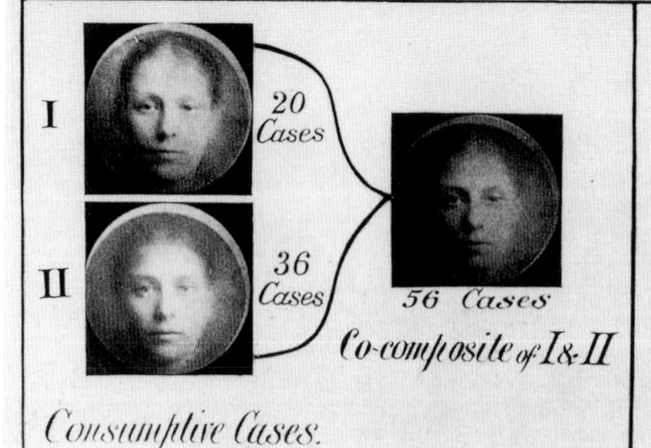

56 Cases

Co-composite of I & II

Consumptive Cases.

100 Cases

50 Cases

Not Consumptive.

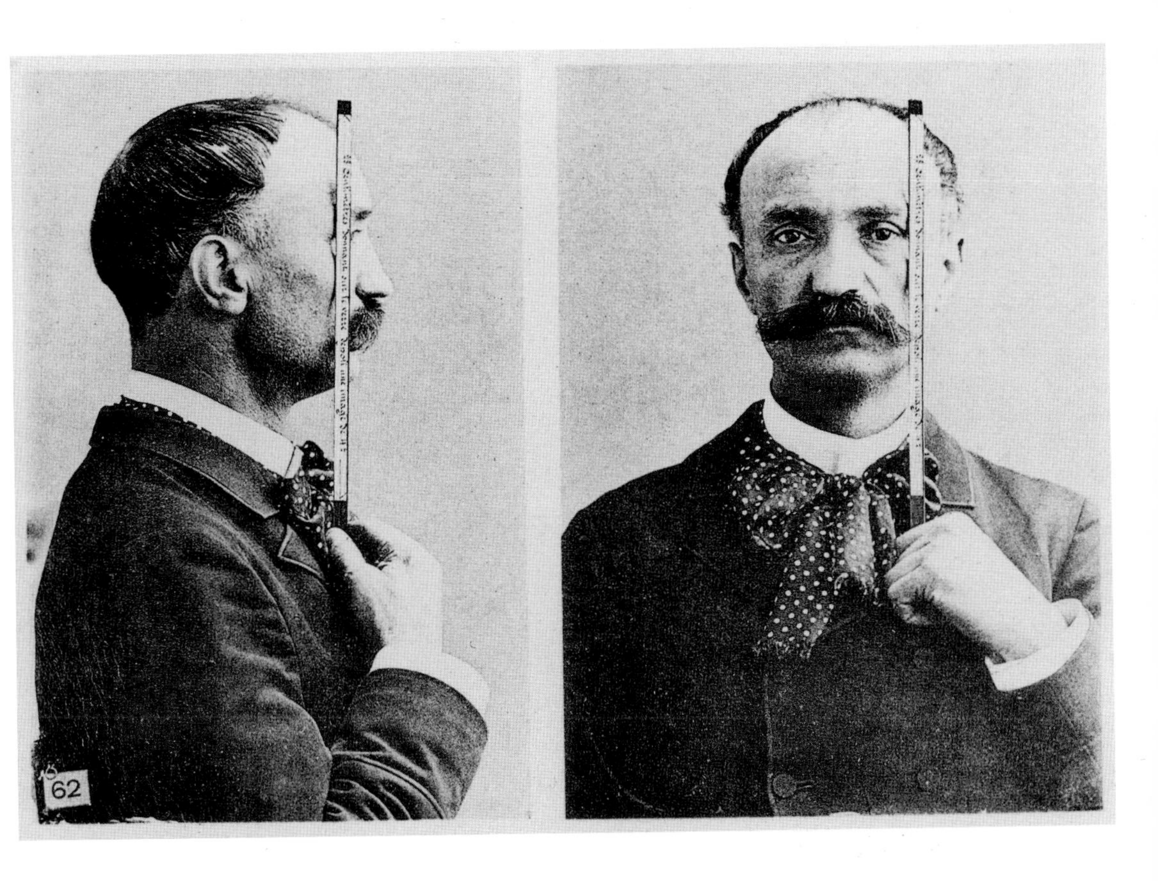
62

1892 short story, 'The Cardboard Box'. Bertillon's detailed photographs of the right ear also appeared in 1933 in *The Phenomenon of Ecstasy* by the surrealist artist Salvador Dali. It is now known, however, that part of the ear continues to grow throughout life, and so the ear may not be as accurate a guide to identification as Bertillon supposed.

mug-shot

As a result of Bertillon's system, the full-face 'ID' photograph has become the usual international method for identifying individuals on passports, driving licences, identity cards and credit cards. It is still the standard symbol for identity in the 21st century. Physiognomy was superseded by the 'hard' sciences in the early twentieth century, but the recognition and classification of facial images is still a subject for debate and one in which technology is playing an increasing role. The police mug-shot is now a familiar image of everyday life and one which features regularly in the press.

Following his arrest in 2003, the mug-shot of Michael Jackson created a sensation. This was due partly to the unmasking of Jackson's 'real' face, ravaged by years of plastic surgery, that the public craved, and partly to the frisson created by the best-selling conjunction of celebrity and crime. Much the same effect had been created two years previously by the mass circulation of mug-shots of the actor Hugh Grant and the prostitute Divine Brown.

The creation of photographic 'stereotypes', another method of anthropological classification, was largely due to the work of the British eugenicist Francis Galton, a leading biological scientist working at about the same time as Bertillon. Galton was fascinated by questions of heredity and human types, 'the biological foundations of human society'. He believed that there was a quantifiable biological hierarchy of human intelligence, and he recommended racial purity. To this end, Galton deployed photographic techniques to analyse and measure different categories of people. From 1865 onwards he conducted a number of experiments to develop his theories.

To do so, Galton created 'composite' photography; that is, if a certain group of individuals shared a particular mental trait and this was somehow reflected physically, the common features might be extracted by superimposing photographs of their faces one upon another. As Peter Hamilton and Roger Hargreaves explain:

> (Galton's) interest in mathematics and statistics had led him to experiments which would offer an 'averaging' of physical characteristics in a single image, so that the 'normal' distribution could be observable in the same way as a graph might indicate how a population's characteristics were related to their statistical mean...[6]

Thus, through the re-photographing of, say, portraits of criminals by successive multiple exposures on the same plate he would create a composite image or a 'photographic mean' of the 'type'.

In 1877, Galton tested his hypothesis by examining thousands of photographs of criminals, obtaining copies of those that interested him. He asked for them to be classified by type of crime rather than by name: first, murder, manslaughter and burglary; second, felony and forgery; and third, sexual crimes. From the resulting 'composites' he drew the conclusion that 'the three groups of criminals contributed in very different proportions to the different physiognomic classes'. His theory was based on the premise that if a distinguishing feature existed 'it would come out in his mixed photographs in a clear line, whereas in those features which do not correspond the lines would be more or less blurred'. His intention was to find a means to identify a tendency towards crime and either eradicate it or find out if the miscreant was 'incurable'.

But composite photography failed to reveal features typifying different groups of criminals, and Galton ultimately admitted the failure of his method, writing in *Inquiries into Human Faculty and Development* (c. 1883) that the composites 'produce faces of mean description, with no villainy written on them'. Galton turned his attention instead to composites of families. In 1882 he sent out a circular letter to amateur photographers, asking them to send him individual photographic portraits of their families. He instructed them to take the photographs in both full-face and profile, under identical controlled conditions. Contributors were offered a copy of Galton's composite portrait of 'the family likeness' in return. Here his intention was to identify heritable facial features and thereby to forecast the physical appearance of the children of a proposed marriage. He later also turned his methods to attempts to define a physiognomy of disease and of mental illness, and to stereotype specific racial types (through composites of Jewish boys). On 25 May 1888, Galton gave a lecture at the Royal Institution in London entitled 'Personal Identification and Description'. 'It is strange that we should not have acquired more power for describing form and personal feeling than we actually possess,' he said.

composite face

Galton's influence in using the face as material for a manipulated picture rather than a penetrative portrait was widespread and continues today. His composites lie behind a number of photographic projects which seek to identify types, such as photographer E. J. Bellocq's private photographs of prostitutes in the legal brothels of New Orleans's Storyville red light district, made around 1912; Walker Evans's photographs of farm labourers of the 1930s for the US government Farm Security Administration; Bill Brandt's 1930s Britain; Nadar's French bourgeoisie; and German photographer August Sander's documentation of German social groups and ethnic identities made in the Weimar Republic during the 1920s. In his introduction to August Sander's famed *Antlitz der Zeit: Sechzig Aufnahmen deutscher Menschen des 20. Jahrhunderts* ('Face of our Time: Sixty Portraits of Twentieth-Century Germans') in 1929, the German novelist Alfred Döblin wrote that Sander's photographs of people were 'photographs of faces', which deployed techniques of scrutiny or clinical observation in which the 'comparative anatomy' of the 'culture, class and economic history of the last thirty years' could be studied. In 1934, the Nazis censored Sander because his 'faces of the period' did not correspond to the Nazi stereotype of the race.

In France, the photographic innovator Arthur Batut further developed Galton's composite methodology during the late nineteenth and early twentieth centuries, using his technique to demonstrate physical characteristics, and clearly stating that his composite portraits were a form of virtual reality or 'images of the invisible'. Specifically, Galton's technique of sandwiching multiple faces to create average or fictive types constitutes the aesthetic foundation for later composite portraits. In January 2000, a computer-generated, multicultural picture of Eve appeared on the cover of *Time* magazine to illustrate the 'face' of America. It was produced by combining the features of Anglo-Saxons (15 per cent), Middle Easterners (17.5 per cent), Africans (17.5 per cent), Asians (7.5 per cent), southern Europeans (35 per cent) and Hispanics (7.5 per cent). The 1993 image of a black Queen Elizabeth in Benetton's *Colors* (no.4: Race) magazine mixed features in such a way as to produce an arresting picture of Queen Elizabeth as a woman of colour, and prompted a lively debate in British newspapers.

A number of artists, including William Wegman, Nancy Burson and Christian Dorley-Brown, have centred their artistic practice on composites. Nancy Burson's composite photograph *Mankind* (1983–5) mocks the taxonomic impulse behind the notion that individuals can be presented and analysed as 'types'. By using advanced scanning techniques to merge imaginary and real faces, Burson's work explores the erasure of identity. By contrast, Dorley-Brown's project *Haverhill 2000* sets out to establish a permanent and accessible image archive of a town's population at a certain moment in time and to create a new 'virtual population' made up of different combinations of faces within six specific age groups. The artist was interested in how the portrait, as a traditional convention in picture-making, can reveal things about individuals, the way they live and how they feel. Using computer software, 2,000 individual photographs were merged and arranged into groups (male and female, grouped 0–5 years, 5–10 years, 10–15, 15–25, 25–40, and 40+; then 0–15 and 15–40+; then 0–40+; and finally a complete Haverhill 2000). This hyper-real, somewhat ghostly and gender-free image is created from photographs of 1,000 females and 1,000 males aged from six months to 80 years old and covering 50 ethnic groupings. With each pairing requiring 40 individual features to be morphed together, the process took over four months. According to art historian and critic Norman Bryson, such manipulated composites prove 'photography's own power to erode the presence of the face and put in its place a cloud, a cathode ghost'.[7]

There is even something of Galton's composites in the portraits of Hollywood stars and in celebrity portraiture: for example, the multiple layering of Warhol's series of silk-screened 'portraits' of movie stars (including Elvis Presley, Elizabeth Taylor and Marilyn Monroe). Celebrity status is more important here than the individuals themselves. And, according to critic John Gross, as the fame of celebrities increases, 'their features evolve into something more than faces – into masks, trademarks, icons'. Or, as the novelist John Updike memorably put it: 'Celebrity is a mask that eats into the face.' At the same time, works like Andy Warhol's screen-printed police photographs of alleged criminals, *Thirteen Most Wanted Men* (1964), and Christian Boltanski's artist's book *Archive* (1989), which juxtaposes anonymous faces of criminals and victims, show the incorporation of the criminal mug-shot into art. Both artists play with the stereotypical visual image of the criminal. The lasting impact of physiognomic and police classification systems is evident in the stereotypes still attached to the faces of criminals. One of the most famous British mug-shots is that of Myra Hindley, imprisoned for life as accomplice to Ian Brady in the murder and torture of a number of children in the notorious 'Moors Murders' of the 1960s. For some, Hindley's face is clearly the incarnation of evil. Others, who tried to secure her release from prison, see behind the blonde hair and direct look the eyes of a victim and a martyr. The way we impose a reading of a face, or cast our images in roles, was perhaps even more apparent in the press reaction to the police photo of another female murderer and torturer, Rosemary West, which was taken thirty years after Hindley's, and is oddly reminiscent of it. As the writer Joan Smith points out, it is perhaps not difficult to construe Hindley's face according to a particular stereotype (bleached blonde, hard as nails, icily sexy). But in the case of West, commentators had a hard time linking the nauseating catalogue of abduction, calculated sexual abuse, mutilation and murder with this plump and homely-looking middle-aged woman, with her badly cut brown hair and old-fashioned glasses. In the words of the *Daily Mirror* feature writer Cheryl Stonehouse: 'See her in the supermarket. See her in the street. See her with her children. Just an ordinary wife and mother.'

The hunger for photographic sensation is not new. Newspapers, especially the tabloids, have always appreciated that it is photography rather than news that sells

Tibor Kalman, **Black Queen Elizabeth**, from Benetton *Colors* magazine, 1993 (no. 4: Race) (p. 120)

Mug-shot of Myra Hindley (p. 121)

Christian Dorley-Brown, **Haverhill 2000** (pp. 122/123)

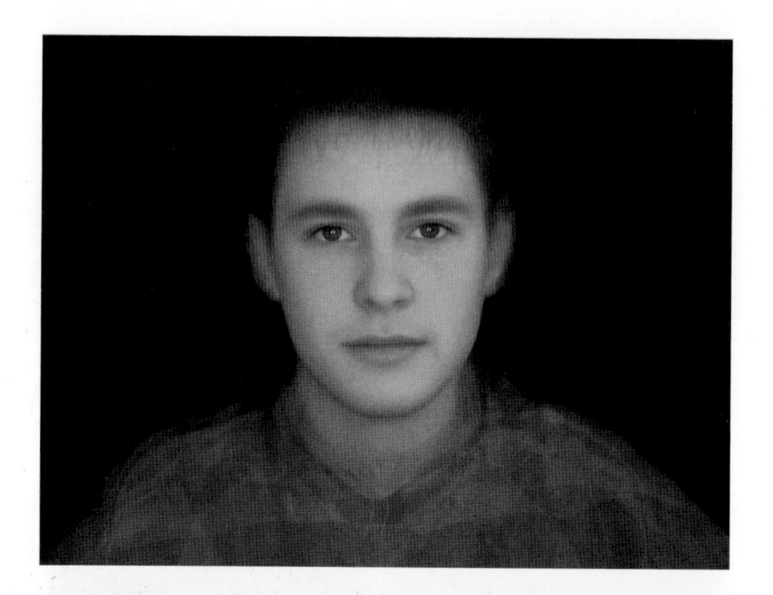

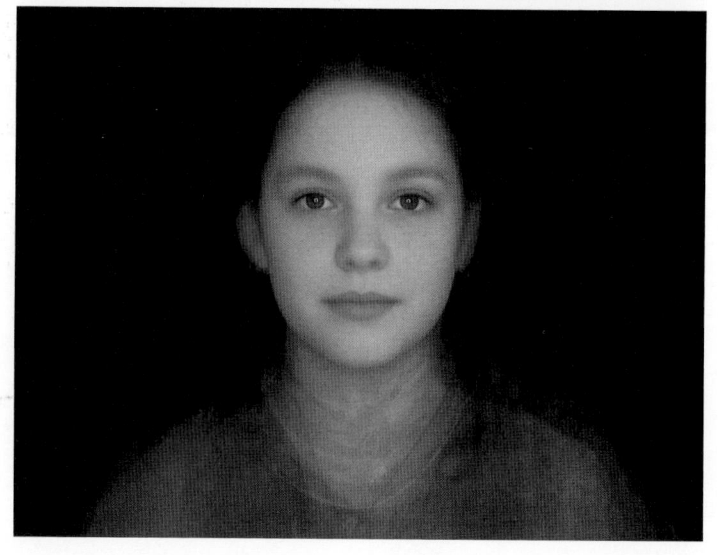

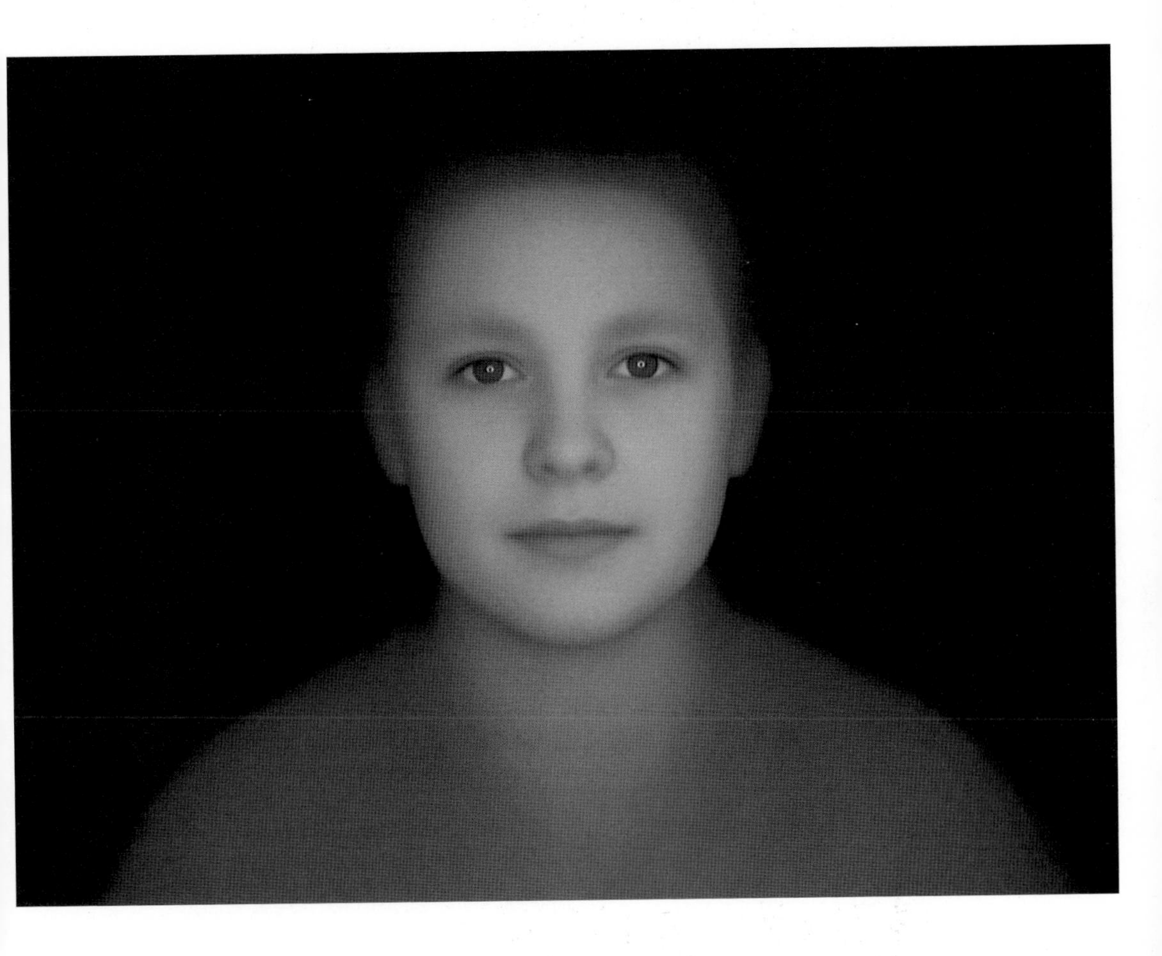

newspapers. From the 1900s onwards, crime and celebrity images have been top of the list. The American tabloid sensation of 1928 was a photograph no one was meant to see: a woman dying in the electric chair. (Her name was Ruth Snyder, she had murdered her husband and her picture appeared the following day under the headline 'Dead!') For a century now, photojournalism has been the dominant mode of profiling identity and a key medium for promoting sensation and celebrity. The development and pursuit of the photographic celebrity image has been powerful, hastening the development of photographic and surveillance technologies.

The cult of celebrity images is ultimately just as dependent on social stereotypes as any of the physiognomic portraits of the last century have been. The late Queen Mother was not unaware of the construction of a particular image of the royal family as iconic through the photographic portraiture of Cecil Beaton. The photographed faces of fashion models on magazine covers are also fabrications. As the supermodel Kate Moss explains:

In a way, it's like the photographer always has his vision of me. The pictures that I am known for are not really my image, they're always the photographer's vision of me. I can look a hundred different ways, but what people see of me in pictures is not really my image.

To mark the Millennium in 2000, *Vogue* magazine asked seven 'Brit' artists – Marc Quinn, Jake and Dinos Chapman, Sarah Morris, Tracey Emin, Gary Hume and Sam Taylor-Wood – to represent Kate Moss however they chose. Taylor-Wood photographed Moss as the original Madonna, Quinn made an ice sculpture of her and Hume sketched Moss from photographs, projected the drawings on her face and took photographs of the overall effect. 'Fame fits the frame,' wrote a *Guardian* leader, reviewing the result: 'Move over Mona Lisa. A hoyden from Croydon will wipe that smile from your face . . . Even a supermodel, if she wants to be remembered, must pause awhile to pose, and not just for the cameraman. She must model for the artist.' Justine Picardie, who commissioned the 'Kate Art', claimed that the choice of artists was particularly apt as they were artists who understand celebrity, having become celebrities themselves.

Two years previously, after the funeral of Diana, Princess of Wales, commentator Serena Mackesy drew attention to a similar addled iconography:

And everywhere you turn . . . the tokens of remembrance building up against the railings have the feel of a Mater Dolorosa offering of a Maltese Easter. Photographs of Diana, crucifixes draped about the sides in rosarial offering. A picture of Marilyn Monroe. Another of the Mona Lisa. Among the piles of lilies and teddy bears, a Minnie Mouse doll. Saint, artist's model or Disney character? Only time will tell.

identity

The development and pursuit of the photographic celebrity image has had a major influence on diverse new technologies. Lucrative industries constructed around both celebrity and surveillance have jointly developed technologies which are now able to produce an amalgamated face by combining advanced biometric techniques, encoding unique facial characteristics with photographic and moving image enhancement. Following Galton's simple technique of superimposition, the use of manipulated and superimposed images of the face and of mass-collection and classification of face data are at the heart of new and virtual face 'capture' technologies. Amalgamated or 'virtual' faces are now high currency because they allow on the one hand applications grounded in the cult of celebrity and the media market; and, on the other, applications grounded in scrutiny (harnessing the face for identification, for data storage, for forensic recognition and classification).

Over the past decade, global terrorism and illegal immigration have increased the drive to introduce ID security technologies based on biometric data. As a result of a global security review, and the growth of identity-related crimes (projected to rob the global economy of $24 billion in 2004), the USA now requires travellers from 27 countries (including Britain) to acquire machine-readable passports containing biometric data relating to facial characteristics, which can then be stored on a national US database and compared with a terrorist 'watch' list. In the next couple of years, details of individuals' physical features – such as facial characteristics, iris patterns and fingerprints – will be incorporated into British identity documents (passports, ID cards and driving licences), and a major trial of biometric technologies linked to these developments is already under way. The British government is also set to press ahead with large-scale trials of universal identity cards, in spite of opposition from civil rights groups.

Biometric security devices generate in-depth 3D facial portraits similar to holograms and secure enough to be embedded in documents. To capture the 3D image, subjects stand in front of a digital camera for three to ten seconds while a projector beams an invisible coded light pattern on to their faces. The scan keys on hard tissue and bone and captures multiple facial angles in minute detail, digitally recording subtle variations to provide accurate information.

While three-dimensional facial recognition technology is in active development, and would be more accurate than fingerprints and two-dimensional mug-shots, there is a big obstacle to its adoption – the lack of existing databases. These have to be built up from scratch. Conversion programs are under way, aimed at taking the huge database of existing two-dimensional images available on identity documents such as driving licences and using the limited information they contain to provide basic biometric data which could be used to verify three-dimensional scans of individuals' faces as they become available.

However, despite advances in the technology, biometrics is by no means 100 per cent accurate. While iris recognition can achieve accuracy of rates just tenths of a per cent short of perfection, the rates for fingerprints and facial recognition are much lower. Of the three, facial recognition is the easiest to fool, using beards or prosthetics, or, in extreme cases, plastic surgery. Work on face perception by psychologists Vicki Bruce and Andy Young at the University of Stirling has also demonstrated that face recognition is part of a whole perceptual framework of sensation and complex mental processes, and this framework needs to be considered in conjunction with any technological reading. Bruce and Young have explored the huge range of experimental data on face processing (on how we recognise faces): why, for example, upside-down faces are so hard to recognise and what makes some faces more memorable than others. They have shown how face recognition is strongly dependent on quasi-artistic effects, the correct balance of relative brightness, and context.

Over the past decade, the growing presence of CCTV in shops, banks and public spaces has become accepted in the courts as a powerful identification tool to back up, or even replace, eyewitness accounts of incidents. According to Vicki Bruce, such confidence is almost certainly misplaced. Changes in viewpoint, expression and lighting (and of course hairstyle and glasses) make recollections of a face notoriously unreliable. Because we have not had time to build up a composite picture of a stranger, we have to rely on the brain's ability to initially perceive objects only in terms of light and shade, particularly shade. Both human vision and computer vision find light and shade hard to deal with. Bruce explains:

Although human vision is quite tolerant of changes in lighting, face perception can be strongly affected by these changes. It is difficult to recognise even very familiar faces when they are lit from below... and it is hard to decide that two faces belong to the same individual when they are lit from very different directions.[8]

Images captured on security cameras are often of extremely poor quality, and thus much interpretation is involved in determining to whom a captured face belongs. New technologies for reading unique facial signatures determined by their underlying vascular structure are currently being developed in North America. State-of-the-art infra-red cameras, in conjunction with image-processing software, could be used to recognise facial signatures.

virtual face

Despite the controversy surrounding biometric technologies, three-dimensional computer animations of heads and faces are no longer unusual: they are used in games and animation cartoons and for medical and forensic purposes. The digital effects industry in film production now frequently integrates animated three-dimensional models into two-dimensional live-action footage, as for example in recent blockbuster films such as *Titanic* (dir. James Cameron, 1997), *Batman & Robin* (dir. Tim Burton, 1989) and *Face/Off* (dir. John Woo, 1997). More recent releases include *Freddy vs. Jason* (dir. Ronny Yu, 2003), *Kill Bill vols. 1 and 2* (dir. Quentin Tarantino, 2003 and 2004) and *The Lord of the Rings* trilogy of films (dir. Peter Jackson, 2001–3) (in which the character of Gollum is a characteristic example).

The invisible man in *The League of Extraordinary Gentlemen* (dir. Steve Norrington, 2003) epitomises the 'seamless' amalgamated actor and data face (or digital prosthesis/augmentation). 'Eyetronics' technology can record an actor's face on film and translate the expressions into a three-dimensional computer model. The system takes as many as 5,000 measurements for each frame of film and is able to track facial features such as the tip of the nose and the corners of the mouth from frame to frame. These movements are then used to animate a virtual actor. The result is such that it is almost impossible to tell the digital and real performances apart.

Software developed for the film industry is also being adapted to model the effect of the different incisions surgeons can make, and to predict how surgery on a person's face will alter their appearance after the operation. Techniques of digital facial reconstruction are also used by archaeologists and forensic scientists to digitally re-animate statistical estimations of dead people's faces from skeletal remains. Three-dimensional graphics programs developed at the Max Planck Institute for Computer Science in Saarbrücken, Germany, by Kolja Kyhler and Jyrg Haber and their colleagues speed up the laborious process of sculpting a face from a skull, but are also able to furnish the reconstructed face with realistic facial expressions. Kyhler and Haber first scan a skull to create a three-dimensional computer model of it. Next, they identify sites on the skull for which tissue depths are available. The software model then automatically adds flesh of appropriate thickness to the skull, suitably adjusted for variables such as ethnicity and sex.

In order to animate the head, researchers have built a generic virtual head that is animated to simulate the facial muscles responsible for expression. Features of the reconstructed skull, such as skull shape and tissue depth, are used to customise the three-dimensional model of a reference-animated head created by the computer as the foundation for reconstruction. The result is a fully animated head that can adapt its expression in the same way as the dead person might have done. 'We can show subtle facial expressions, and

Biometric mesh (p. 128)

Heather Barnett, **One man's land**, 2002 (p. 129)

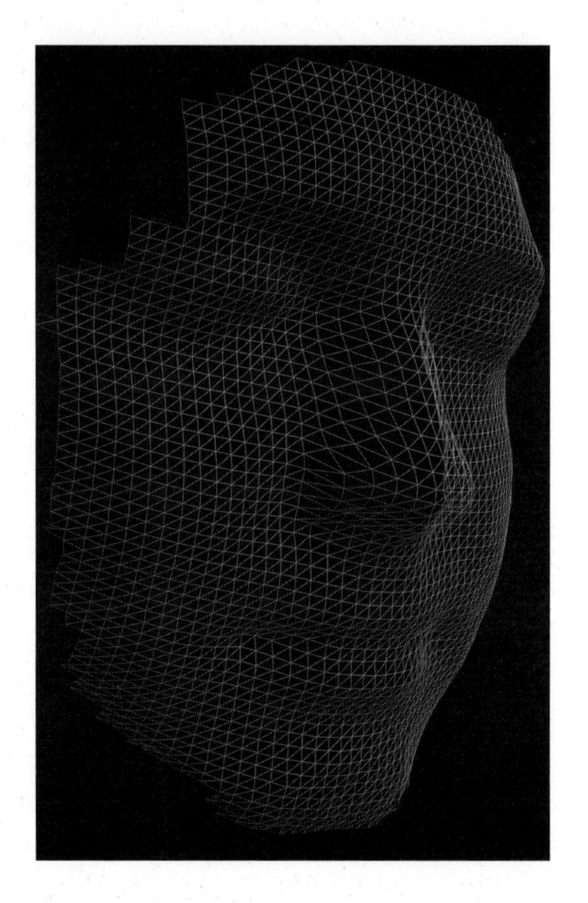

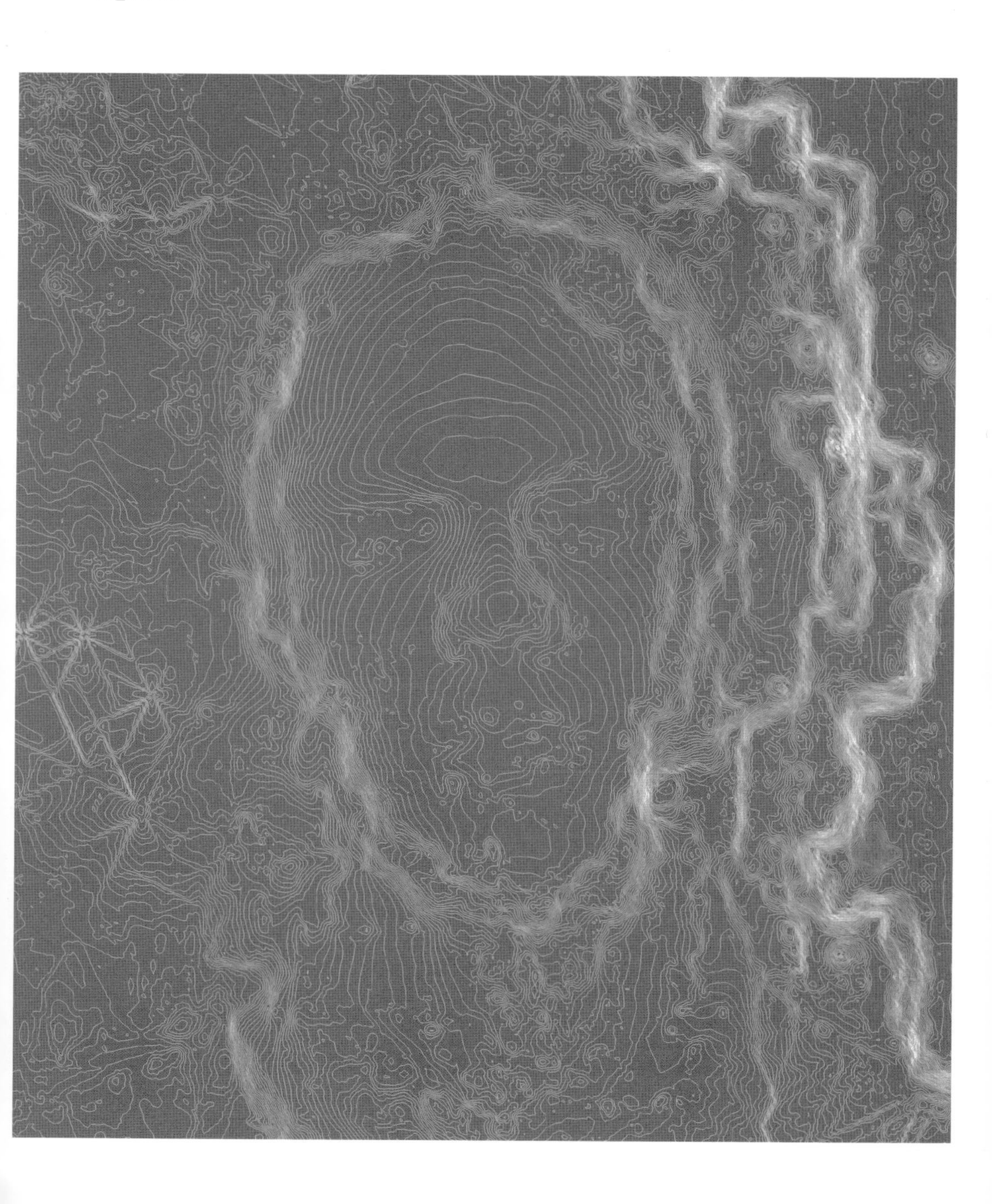

that's something that the traditional clay sculpting method cannot do,' says Haber. For instance, creating a friendly or angry face with traditional methods would mean going back to the sculpting table. 'With our system, it's just a few clicks of the mouse, and you can change the expression.'

Similar technologies can now also create virtual clones and (digitally) resurrect the dead. The hauntingly beautiful *Digital Marlene* was created by digital animator Daniel Robichaud for Virtual Celebrity Productions, now the licensing and merchandising company Global Icons, which aims to 'help celebrities protect and extend themselves as brands'. An increasing number of new digital cloning systems allow resurrected celebrity faces to be grafted on to contemporary actors' bodies: by using the movements of a live actor to animate a digital look-alike, the dead celebrity can have a posthumous acting career. There are, of course, also significant benefits for the actor's estate and licensing company. Some relatives may well object, but according to Dietrich's grandson the task for the heirs of deceased movie stars is:

> to take these iconic performers, maintain their essences to the best of our abilities . . . then find the new Picassos, the new creative people with the tools to take this art form to the next level. It's our job as their offspring to maintain the flow of their creativity.[9]

Sony, Pixar, Industrial Light and Magic and other major visual entertainment companies have already produced digital actors with key roles in feature films. Among the best examples are the mouse that stars in *Stuart Little*, the insects in *A Bug's Life* and Jar Jar Binks in *Star Wars: Episode One – The Phantom Menace*. Creating digital humans is more difficult. While they have been used in crowds and as virtual stunt doubles, they wouldn't stand up to close scrutiny, so none has yet played a role requiring an extreme and/or sustained close-up. 'Everybody looks at [humans] every single day, and even the untrained eye can tell when it's not right,' says Derald Hunt from Kleiser Walczak, a special effects firm that is a pioneer in the digital-actor field. The most difficult thing to reproduce is muscle movement, especially in the face, where the movements of muscles are extremely specific. Even shading an eyeball so that it reflects the proper amount of light is a challenge yet to be solved.

Virtual clones are created using the same basic technology as digital resurrection, but these appear more 'real' or believable, as they are generated using high-resolution three-dimensional scans of live faces and are not limited to reanimation of existing two-dimensional material. Virtual Clones (the company developing this technology) uses three-dimensional photography to first capture three-dimensional models and then process them to create an accurate 'depth map' of the chosen subject. The depth map is then converted to a polygonal model and rendered with high-resolution texture maps. In essence, the technology captures the complex layers of the face's physiognomy and maps them on to a three-dimensional high-resolution scan of the face, using other software for expression, eye-movement, voice and emotional tracking. In theory, such techniques could be used to create an independent version of another person that – as data – could be easily kidnapped and put into different arenas without consent. The misappropriation of such images is a major problem in the digital age.

ownership

The issue of facial ownership is tantalisingly unclear, both legally and morally. Who owns an image or representation of a face? Is it possible to distinguish between representation and ownership (or privacy rights) of public or private faces? How are distinctions between use and ownership drawn?

In 1998, the estates of Elvis Presley and Diana, Princess of Wales, were both unsuccessful in court bids to patent the deceased's faces. In the first case, brought by the estate of Elvis Presley against an East London memorabilia shop called 'Elvisly Yours', the judge ruled that 'there is nothing akin to copyright in a name' and 'similarly, Elvis did not own his own appearance'.

In the months immediately after the death of Diana, Princess of Wales, in 1997, her estate fought a losing battle to protect her image from a flood of tacky memorabilia and unauthorised books, including T-shirts, mugs, plates, 'antique bronzed medals', stamps, and even, on a busy road outside Tel Aviv, a twelve-foot-high portrait of Diana with the caption 'Drive Carefully', painted by the Israeli artist Avi Nahmias. For the past twenty years, Diana's face has been reproduced more than any other in the world. As journalist and author Nicci Gerrard wrote in the *Observer*'s commemorative edition, a week after Diana's death:

> Her life was meant to be watched, not heard. She was an image which outdid all others. We followed her life, frame by frame: the sweet princess reeling through the years, fast-forward and pause. Everywhere we looked, she looked back at us, from magazines, newspapers, television screens, posters.

Even commentators usually at loggerheads agreed that Diana had been 'the most photographed and debated woman on earth ... the most cherished icon of the modern era' (Andrew Morton, Diana's biographer) and that 'the degree of her celebrity eclipsed even global superstars such as Marilyn Monroe and Elvis Presley' (Joan Smith). When lawyers from The Diana, Princess of Wales Memorial Fund sought to establish Princess Diana's face as a trademark, 26 pictures of her which 'sum up her life in the public eye, capturing her unique charm and appeal' were sent to be registered. ('Bid to patent Diana's face', ran the headline in the *Daily Mail*.) When the attempt failed, Diana's estate then attempted to patent a virtual Diana made out of a composite of the

Pierre Huyghe and Philippe Parreno, **Skin of Light**, from the series 'No Ghost just a Shell: Annlee' (p. 132)

Daniel Robichaud/Global Icons, **Digital Marlene** (p. 133)

26 images, but this too was turned down by the courts.

There is an interesting analogy here with the ongoing dispute over the Rice portrait of Jane Austen. As a famous eighteenth-century novelist, Jane Austen has the status of a British national treasure, and the Rice portrait has been accepted until recently as the only reliable existing image of her. Now, however, some art historians are disputing that this image is, in fact, Jane Austen's. The resulting furore raises interesting questions about who is authorised to speak for her, and to whom she 'belongs'. Reviewing the debate about the authenticity of the iconic image, critic Claudia Johnson asks:

How could any image be commensurate to what we think and feel about Jane Austen? How could so momentous and yet so intimate a figure, one who has become the site of such different and such contradictory visions and fantasies – about civility, love, history and England itself – survive figuration without arousing disappointment or anger?[10]

Marcus Harvey's *Myra* is another fascinating example of this controversy. *Myra* reworks one of the most memorable photos of our time, the tabloids' favourite Myra Hindley photo, by using a template of a child-sized hand-print hundreds of times to build up an enormous painting measuring 9ft by 11ft. Hindley herself protested from prison about the use of her face on this image and its alleged effects on her applications for parole. In 1997 the Vice Squad was called to the Royal Academy's 'Sensation' exhibition to consider 'a reasonable basis for prosecution' with respect to the portrait; none was found. When *Myra* was subsequently attacked and defaced, the police were again called in, this time to protect the portrait.

Computer-game heroine Lara Croft, from the best-selling 'Tomb Raider' series, is a more recent example of issues relating to the ownership and marketing of the virtual face. She has appeared on the covers of mainstream magazines such as *The Face*, *Melody Maker* and *FHM*. On the one hand, *Playboy* magazine was banned by the High Court from featuring Lara Croft as a cover girl – much to the disappointment of the 'human incarnation' of the 'Tomb Raider' heroine, model Nell McAndrew, who spent a year doing promotional work for the game.

The High Court ruling followed claims by Core Design Ltd, owners of the Tomb Raider trademark, that Lara Croft's 'squeaky clean' image could be 'tarnished for all time' by association with *Playboy*. On the other hand, Science Minister Lord Sainsbury used Lara Croft to promote UK science and technology with Core Design's consent. 'I want Lara Croft to be an ambassador for British scientific excellence,' Sainsbury said. He did, however, acknowledge that the provocativeness of this image might be less than appropriate to represent the best in British research and development.

avatar

Lara Croft is also an example of an extremely successful games 'avatar' – so successful that she was brought to life in a film version. 'Avatar' is a term from computer gaming for a figure or image used to represent a person. It was popularised in the novel *Snow Crash* by Neal Stephenson, but originates in Hindu mythology as the name of the temporary body that a god uses when visiting Earth. Avatars permit role-play and interaction with other people online. They offer opportunities for self-expression and social interaction, with subtleties and complexities not seen in text-only chat-rooms (where the only means of communication is via typed text appearing in balloons that pop out of your head or body – as in comic strips). Avatars are designed to incorporate a wide range of behavioural and emotional expression. They may be created either entirely as computer graphics, or by mapping digital photographs of a real human being on to a virtual three-dimensional model. Unlike entertainment avatars, however, ordinary online avatars tend to be low-resolution animations because of the need for simple and fast processing in a 'conversation' or interface situation. With developing software, however, it is likely that it will be soon be possible to create a photo-realistic avatar of an individual online, incorporating expressions and emotions.

For the most part, however, 'new' and virtual faces are created as devices or technological platforms. Kaya is a digital model, created by Alceu M. Baptistão, a Brazilian animator and special effects artist who intends to launch her as a virtual star. She is one of the most technologically developed examples of a new generation of digital models who have begun to appear on television adverts, in films and on mobile phones. A computer-generated character called Ananova, modelled on a mixture of Victoria Beckham, Kylie Minogue and Carol Vorderman, became Britain's first virtual newscaster from Britain's Press Association, launched on the internet in 2000 and also available via mobile networks and other digital services. Her makers, PA New Media, explained that she had the characteristics of a single 28-year-old 'girl about town' who loves Oasis and the Simpsons and is 5ft 8in tall. She is fully animated, using the latest in three-dimensional computer graphics, and is programmed to deliver the news in a 'pleasant, quietly intelligent manner'. Ananova is the approachable interface for the computer's capacity to produce news bulletins on demand and at superhuman speed, at any minute of the day. In Japan, digital models have been used to sell anything from computers to cosmetics to cash loans. Idoru, the web-roaming cyberstar in William Gibson's 1996 novel of that name, was inspired by the world's first virtual pop star, Kyoko Date, who was created by the Japanese talent agency HoriPro, and who hit the charts in Tokyo in 1996.

Fine artists are also exploring the possibilities of digital models. In 1999, artists Philippe Parreno and Pierre Huyghe bought the image file and copyright of Annlee, a digital model from the Japanese firm Kwork, which specialises in the development of characters for manga cartoon animation and computer games. Annlee was cheap – she cost only 46,000 yen (about £226) – and her facial features and personality were only minimally sketched out. She was more of a tool or a platform than a person. More complex figures are more expensive because of the time invested in the construction of the dynamic mix of properties that forms them. Huyghe and Parreno then 'open-sourced' the Annlee file, inviting fellow-artists to develop her further:

Work with her, in a real story, translate her capabilities into psychological traits, lend her a character, a text, a denunciation, and address to the Court a trial in her defence . . . Do all that you can so that this character lives different stories and experiences.[11]

human face

While the human face is a universally recognised vehicle for the demonstration of technological advances, work on the face as an 'interface' for the future seems to be intent on turning new sophisticated types of data back into an appearance of human tissue based on standard ideas of beauty.

In this respect, Kaya is highly unusual due to Baptistão's decision deliberately to model in imperfections. What marks Kaya out from earlier models such as Lara Croft is the inclusion of facial blemishes. She has freckles, bushy eyebrows and slightly chipped teeth. By contrast, the rest of the industry appears to be geared to creating simulations that are as 'perfect' as possible. While striking as images, however, these avatars have a sense of lifelessness when animated, particularly with respect to the eyes and the plasticity of the face. Despite the success of films like *ET*, *Toy Story*, *Shrek*, *The Wrong Trousers*, *Star Wars* and *Lord of the Rings*, all of which demonstrate that convincing and believable 'cartoon' facial animation is now possible, it is interesting that the makers of Lara Croft felt the need to give her substance through recasting the digital character in human form.

In *Computer Facial Animation*, a pioneering textbook by computer animation researchers Keith Waters and Frederick Parke, the authors argue that faithful human-like computerised faces do not yet exist. In their view, this is because facial animation based on an anatomical model incorporating the complex interplay between bones, cartilage, muscles, nerves, blood vessels, subcutaneous fat, connective tissue, skin and hair has not yet been developed. While research in this area is moving fast and the complete animation of faces in terms of muscle structure is not far off, a more long-term problem is the production of a digital person/face that both seems true to life and can converse interactively with human beings. Eye gaze and movement present a particularly complex challenge to animators, as different patterns are required depending on whether the character is receiving (listening) or giving (talking) information. Movement such as a blink, smile or raised eyebrow can convey diverse and often subtle feelings – this is crucial in the replication of human communication. Spontaneous reactions to questions or remarks help create a convincing connection between a virtual human and a person.

Researchers at the Beckman Institute for Advanced Science and Technology at the University of Illinois are now building up databases of facial expressions from people of different ages and from different gender and ethnic groups which can be generalised and reapplied to virtual faces. It is anticipated that virtual faces could play a key future role in the diagnosis and treatment of patients with mental disorders. Facial photographs have long been used in clinical and research psychology to help understand and map brain function. Photorealistic three-dimensional animated faces could be used to create virtual patients for training and interviewing.

Artists have also been quick to deploy and experiment with the digitised facial image. Cultural theorist Mark Hansen has examined a number of these artworks in the context of the human–computer interface, exploring in particular French-Canadian interactive video artist Luc Courchesne's *Portrait no. 1* (1990) and the German-born digital artist Kirsten Geisler's *Dream of Beauty* (1999). In each installation, the viewer is invited to speak with a virtual figure/being whose face is presented in close-up. In *Portrait no. 1* the female virtual persona is called Marie, and the viewer can select one of six languages to engage in a question-and-answer session with her. But Marie's conversational repertoire is strictly limited and the viewer soon bores of trying to converse with her. 'It is true that I am unreachable and that you cannot change me,' she acknowledges. 'But look at the people around you: Are they so different from me? Are they reachable?' As Hansen points out, interaction with Marie forces a recognition of our intense desire to engage emotionally with the virtual while at the same time it presents the virtual's utter indifference to us.[12]

Like Courchesne's Marie, Geisler's *Dream* seems to offer the viewer the possibility of interactions with a giant, digitally generated, close-up image of an attractive female face. But she is as impervious to intimacy as Marie. Conversation with her provokes no more than the occasional movement of the head, a laugh, a smile, a quick blown kiss. The more the viewer attempts to prompt a response, the more the encounter with the *Dream* face is unsatisfactory. (The more so perhaps because we are now accustomed to the heightened expressive qualities of the face in close-up in cinematic film.) Facial signals usually trigger a sensuous response, but this dialogue or encounter lacks emotions or expression and is disembodied.

Alceu Baptistão, **Kaya wireframe** (p. 138)
Alceu Baptistão, **Kaya** (p. 139)

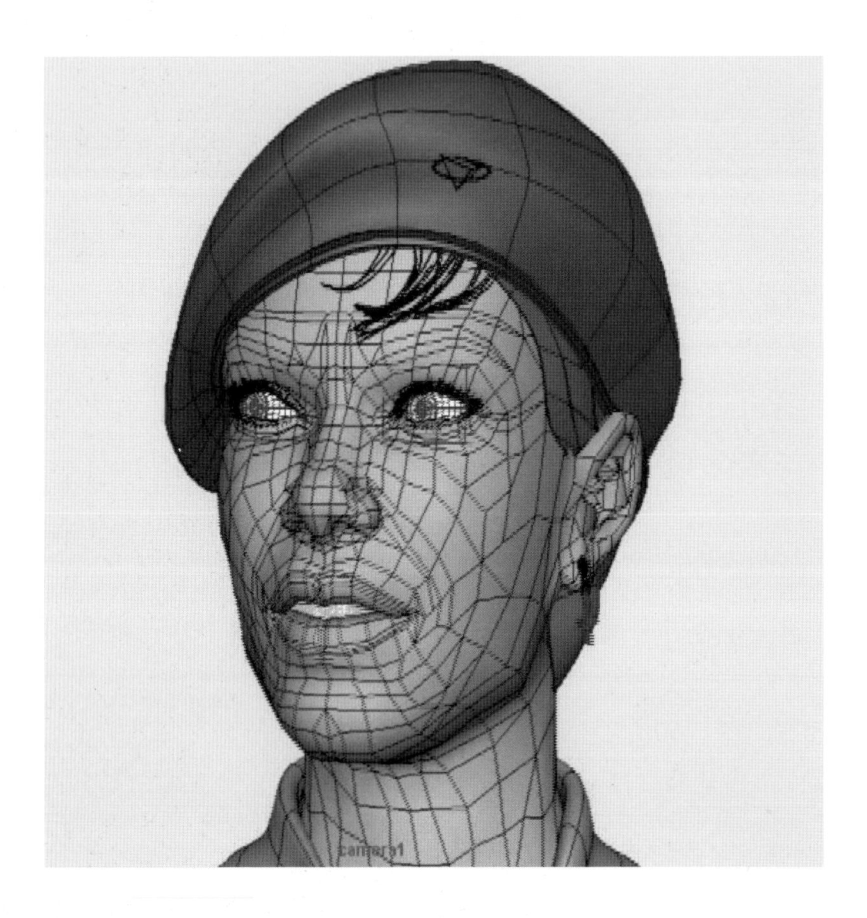
camera1

Meaningful communication or experiential contact with the digital appears impossible.

Another artwork, *Touch Me*, by Alba d'Urbano, an Italian-born artist practising in Germany, also explores the tenuous relationship between the viewer and the digital facial image. In this work, the viewer engages with a close-up image of a woman's face on a monitor positioned at eye level. When the viewer touches the monitor, the image disintegrates. As the screen clears, a live video of the viewer appears on the screen in the place of the woman's face, which then returns to replace the video of the viewer.

In all three of these interactive installations the viewer is marginal, even irrelevant, to the existence of the virtual. As these works demonstrate, the digitised image is not just a copy or representation (like a photograph or a film). It is a simulation that is neither drawn nor recorded but calculated and programmed, and that leads an independent 'life' of its own. It has a potential for autonomous replication and transformation not previously possible technologically. It also emphasises the already uncertain boundaries between the 'real' face and the image, and the animate and the inanimate. In Alexa Wright and Alf Linney's interactive computer installation *Alter Ego* (2004) the viewer is invited to sit in front of a mirror and interact with their own reflection. After about thirty seconds, however, the reflection begins to respond or react. For example, if the viewer smiles, the virtual face or 'alter ego' may look surprised or angry, or may smile back. As Catherine Ikam remarks of the virtual portraits she has been working on since 1995, in these works the face is no longer a 'record' but a 'process':

The work of art is no longer dependent upon a precise medium to which it is linked once and for all: it can now be considered in a variety of ways. Art now involves not so much the creation of artefacts as that of open fluid systems that are amenable to change. The interactive work of art is not an object but a process whose geometry is defined both by its maker and by those who explore it, creating their own perspectives.[13]

Tales about creating forms of artificial life are to be found in the literature and mythology of all cultures and are evidence of a compulsive desire to bring our creations to life. Ovid's account of Pygmalion begging the gods to breathe life into his statues epitomises this longing. For the same reason, ancient cultures were also fascinated by human and animal automatons.

More recently, robots, androids and smart computers have taken on the attributes of living beings. As many examples from literature and film demonstrate, however, there remains a profound divide between artificial life and our own. 'As a general principle,' wrote Oscar Wilde, 'Life imitates Art far more than Art imitates Life.'

Man Ray, **Marquise Casati**, 1922 (pp. 142/143)

Paradoxically, when looking at a dead body or a dead animal, the human brain 'reads' it as related to the living – we know it is dead, but it is from life. By contrast, however 'lifelike' a robot or avatar, the brain does not (or cannot) perceive it as alive. What is still lacking in the collision between the organic and the technological is the 'experience' of a digital face, or a bodily or sensory response to the image. Without these, the digitised expressive elements fail to communicate. The face is powerful in this respect: as an index of the heart, it has so far proved impossible to replicate digitally.

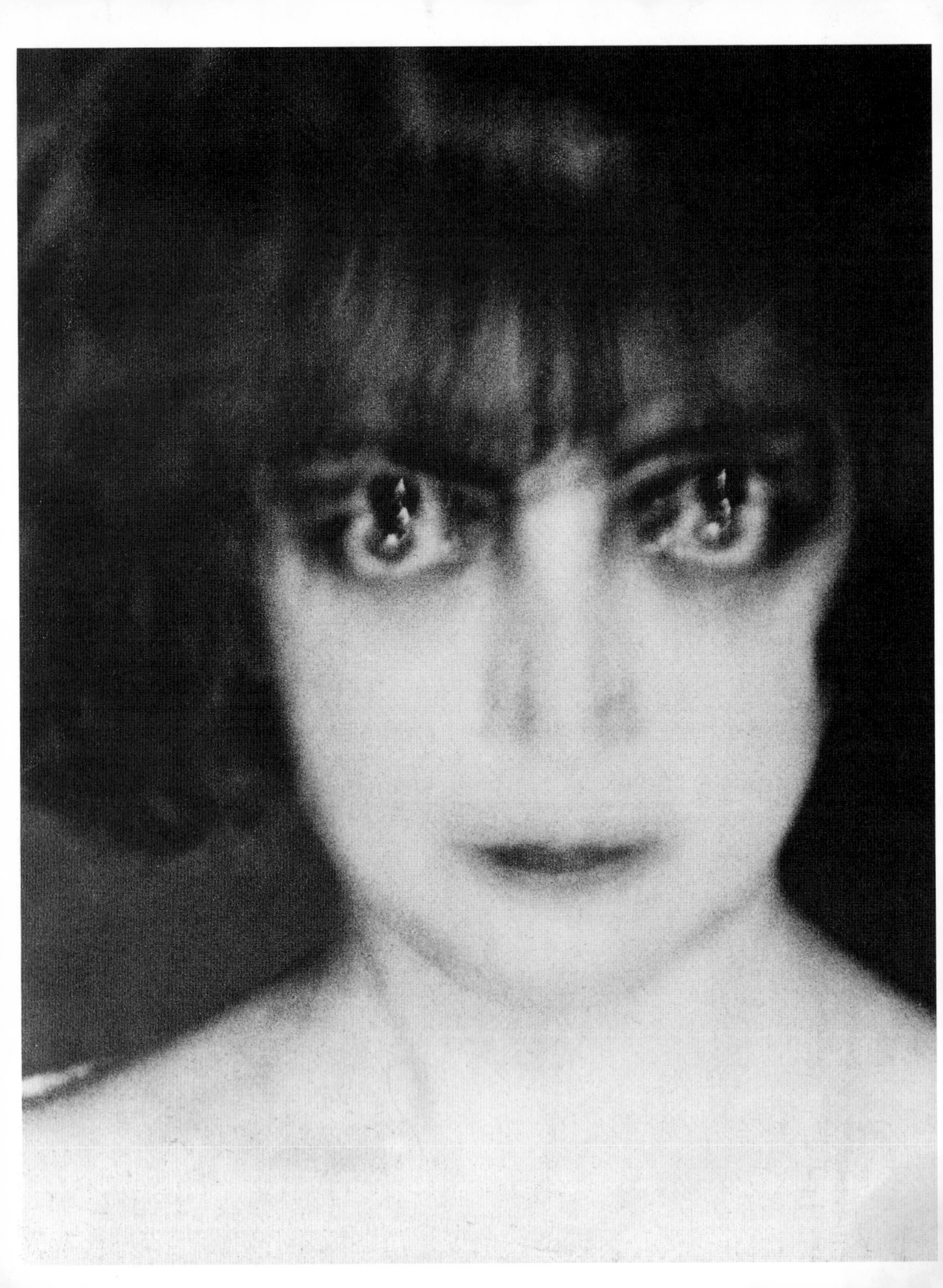

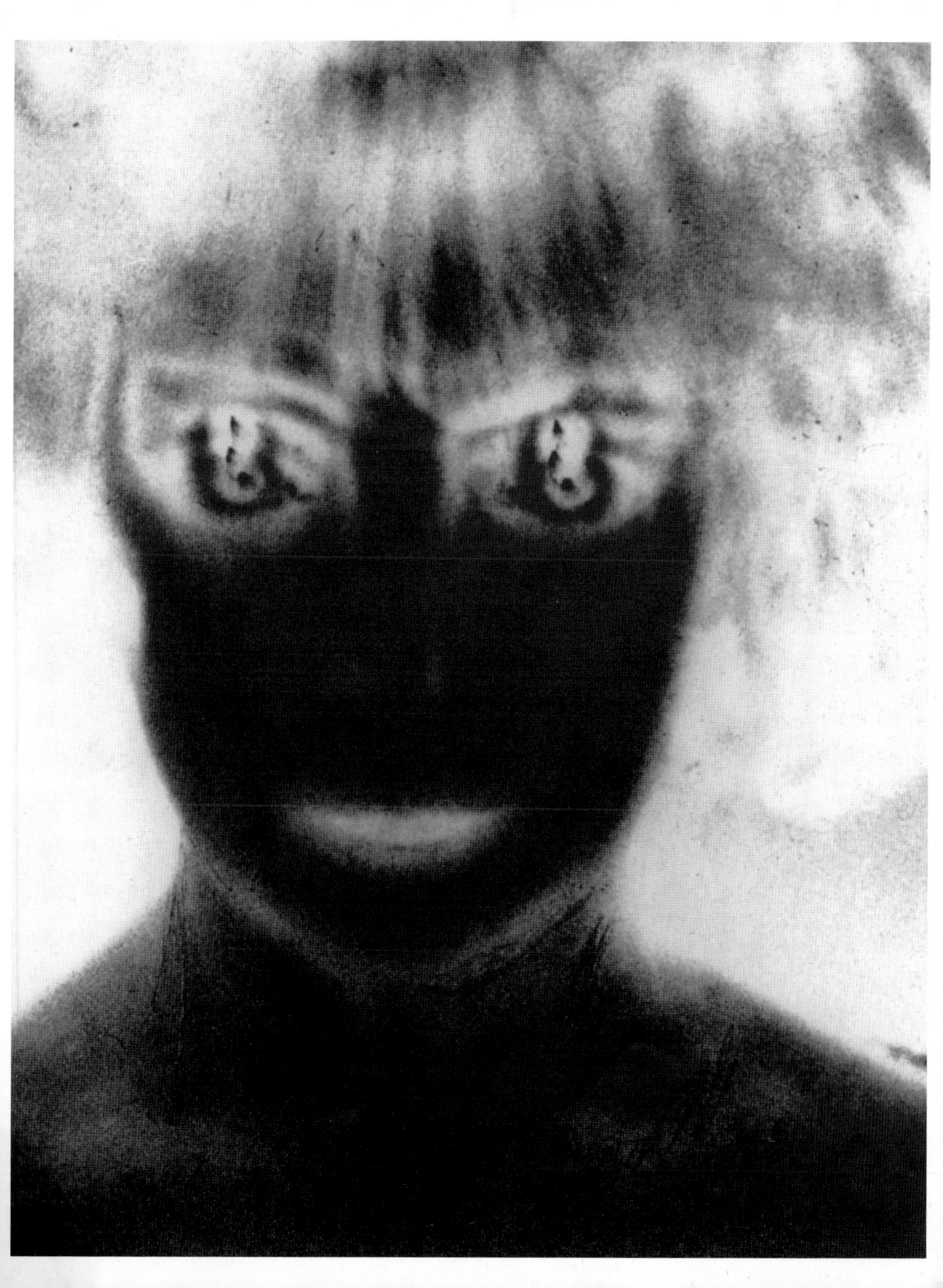

The street was empty; its emptiness had gotten bored and pulled my steps out from under my feet and clattered around in them, all over the street, as if they were wooden clogs. The woman sat up, frightened, she pulled out of herself too quickly, too violently, so that her face was in her two hands. I could see it lying there: its hollow form. It cost me an indescribable effort to stay with those two hands, not to look at what had been torn out of them. I shuddered to see a face from the inside, but I was much more afraid of that bare flayed head waiting there faceless.

Rainer Maria Rilke, *The Notebooks of Malte Laurids Brigge*[1]

extreme face

Our responsiveness to the expressive mobility of the human face has been diminished by over-exposure. The ubiquitous media face offers a remote and retouched idea of the human and fails to move us. Disturbingly, over-exposure to the faces of the dead and dying in newspapers and on television is beginning to have the same distancing effect.

However, the monstrous, 'faceless' or unrecognisable face provokes an immediate sensory response. The suggestion of contact and contamination, the exposure of the visceral and the fear of the unknown and the uncontrollable, all contribute to the magnetism of the grotesque.

In conclusion, this chapter offers a sequence of images or vignettes of the distorted or the disfigured face. These are intended to be suggestive rather than exhaustive and are offered as snapshots, or provocations to thought. Perhaps it is at the margins or boundaries of the face that we come closest to its meaning and power. At one extreme, films like Jonathan Demme's *Silence of the Lambs* (1991) and David Lynch's *Lost Highway* (1997)

Wanda Wulz, **Cat and I**, 1932

Douglas Gordon, **Monster I**, 1996–7 (pp. 148/149)

have used the superimposed, lifted and distorted face to terrifying effect. In instances like these, it is not just the horror at what we see that shocks us, but the realisation of the tenuousness of our own humanity and the fragile boundaries between the human and the animal, the animate and the inanimate.

monstrosity

Monstrosity is as intimately linked to our concept of the human as beauty, and both have accrued moral attributes. The influence of the eighteenth-century moral philosopher Immanuel Kant, who believed that our response to beauty (and to its opposite) is unmediated and personal, is still dominant today. However, there is powerful evidence of the contrary: that it is contextual and subtly connected with social values and with moral attitudes and beliefs.

In the late eighteenth century, Johann Kaspar Lavater had linked beauty to virtue and ugliness to its opposite:

Morally deformed states of mind have deformed expressions; consequently, incessantly repeated, they stamp durable features of deformity ... The morally best, the most beautiful. The morally worst, the most deformed.[2]

Lavater uses the image of the Greek god Apollo (specifically, the statue known as the Apollo Belvedere) as his model of beauty, and associates bestial and animal images with depravity and evil (as the opposite). In his ground-breaking physiognomic text *Physiognomische Fragmente zur Beförderung der Menschenkenntnis und Menschenliebe*, a frog's profile is morphed into a human profile through a sequence of twenty-four drawings. The last three faces resemble an Apollo, wearing a laurel wreath. Lavater was influenced by the Dutch anatomist Petrus Camper, the first to depict the evolution of an Apollo from a monkey in his illustrated lectures, posthumously published in 1791. Morphed similarities between human and animal faces and personalities were featured as early as 1586, *in De Humana Physiognomonia*, by the scientist Giambattista della Porta. In the ancient world, examples can be found in Petronius's *Satyricon*, Ovid's *Metamorphoses* and Apuleius's *The Golden Ass*. The medieval world was full of representations of distortions of face and figure and conflations of human and animal characteristics. These are epitomised in medieval *danses macabres* and Dante's *Inferno*. In the nineteenth century cartoon treatments of ethnic minorities often conflated the human and the ape. For example in *The Tomahawk* of 18 December 1869, a cartoon entitled 'The Irish Frankenstein' by Matt Morgan shows a human body with a large gorilla head, and the word 'Fenian' scrawled across its chest. Today, traces of these representations remain in political and celebrity caricatures with their comic exaggerations and parodic rearrangements of features.[3]

Lavater's equation of beauty with virtue and ugliness with vice set the pattern for the face-reading 'science' of physiognomy for more than a century, and even today such thoughts may underlie our obsession with beauty and the search for the perfect 21st-century face. Yet we are equally drawn to its opposite: oddness, deformity, monstrosity. Mythology and legend are also packed with facial damage: the faces of witches, gargoyles and and trolls, the face marked by leprosy or plague and more subtly, perhaps, the mark of the scar.

The most ingenious and advanced digital technologies now serve to make monsters visible. Anthony Vidler, Professor of Art History and Architecture, University of California, Los Angeles, has explored many aspects of the new transformed spaces created by digital technologies. Is this merely fascination and titillation on our part, a response to the same impulses that drew the Victorians in crowds to 'freak shows' such as the display of the grossly deformed 'Elephant Man', John Merrick, an incurable victim of von Recklinghausen's disease, in a circus booth? Or can monstrosity be used to understand, interrogate and comment on the face? (The verb 'monstrare' means 'to show, teach, inform', and 'monstrum' means something 'wondrous, unsayable, unrecognisable'.) The poet W. H. Auden argued that it can: 'the oddity of the human animal expresses itself through the grotesque'.[4]

Johann Kaspar Lavater, **From frog to Apollo**, circa 1855, from Lavater's *Essays on Physiognomy* (p. 152)

distortion

The mutations and transformations of face and body (particularly in animal-cum-human hybrids) have long been fixtures in culture and art, and are a cornerstone of cinema. An early example is the Medusa of ancient myth – originally a Libyan serpent-goddess worshipped by the Amazons, she is a characteristic image of beauty distorted. Her punishment for being too beautiful (for a mortal) was to be made grotesque through the transformation of her hair into snakes. In some versions of her story, she is killed by a mirror reflecting her own self-image back to her:

She had once been a maiden whose hair was her chief glory, but as she dared to vie in beauty with Minerva, the goddess deprived her of her charms and changed her ringlets into hissing serpents. She became a monster of so frightful an aspect that no living thing could behold her without being turned into stone.[5]

The werewolf, or wolf-man, is perhaps the most striking mythological intermingling of animal and man. The Romans called him 'versipellis', or pelt-changer. The condition was traditionally believed to be caused by a severe sickness and manifested through the wearing of a wolf's costume. Later, in nineteenth-century literature and twentieth-century films, a literal metamorphosis was thought to take place, in which terrifying physiological changes were actually observed, affecting bone structure, teeth

Giambattista della Porta, **Biggest head**, 1586 (p. 153)

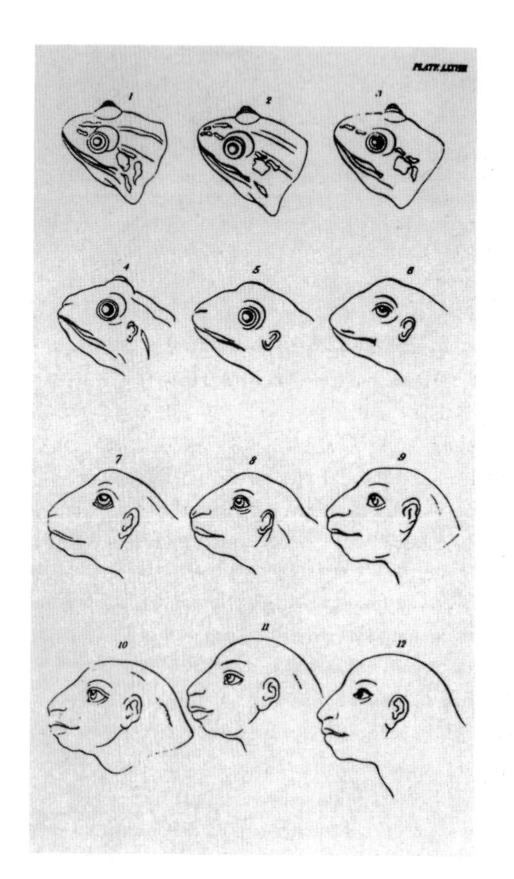

and the skin on face and body. Werewolf images are now the staple of a certain kind of cinema, in which the moment of transmutation and metamorphosis is key. The fear is of two or more identities which alternately possess the body and of the violation of the categorical distinction between humans and animals. The Internet Movie Database currently lists more than 600 werewolf-related films, including *An American Werewolf in London*, *I Was a Teenage Werewolf* and many others. Audience horror at the idea of broken or erupted facial skin is linked with the fear of contact and contamination. This is taken to its extreme representation in the zombies of horror films, whose bite literally contaminates the bitten with the zombie, undead state.

The Medusa and the werewolf faces are both more powerful in their hybrid animal-cum-human form than more recent attempts to imagine future 'alien' faces. Generally characterised by asymmetry and extensions or distortions of a recognisable human face, the 'alien' face typically features enlarged eyes or brain, bulges in the forehead, extended ear-tips and lobes, shrivelled jaw and re-styled hair. In the first decades of the twentieth century, both artists and plastic or 'aesthetic' surgeons wanted to redesign or remodel the face. Aesthetic surgeons were restructuring faces to conform to classical images of male and female beauty, and their before-and-after photographs, using the faces of statues from classical antiquity, look astonishing to us today. By contrast, many artists (Cézanne, Modigliani, Gaudier-Brzeska, Picasso and Matisse, for example) were reconfiguring the face through the disfiguration and even obliteration of facial features. Writing of Matisse's *Jeanette 1-V* (1910–13), the critic Erin O'Connor, in the context of the above, remarks on her progressively deformed features:

Over time the nose lengthens and broadens, losing its bone structure as it becomes a hunk of tubular clay. The mouth flattens and stretches; no longer cut with the clarity of lips, its edges blur into the surrounding material. The hair morphs into two hard knobs and then simply disappears. And the forehead bulges into a blob that blends into the bulky ex-nose. The eyes are particularly telling, widening and separating so far that they are finally only visible as suggestions of eyes, as abstracted, abstracting studies in volume, arc and line . . . No longer representations of the window to the soul . . .[6]

Like *Jeanette* and Picasso's *Female Head* (1932), many of the heads of early abstract sculpture were fragmented and contorted. Rodin's controversial sculpture, *Man with a Broken Nose* (1864) was not exhibited at the Paris Salon as it was considered a travesty of form. The fragmented faces of early twentieth-century sculpture appear in opposition to the ideal classical beauty promoted by aesthetic surgeons.

Cinema, with its dependence upon and exploitation of the face, has a complex relationship with the face in close-up. 'The possibility of drawing near to the human face', wrote Ingmar Bergman, 'is the primary originality and distinctive quality of the cinema.' The philosopher Gilles Deleuze, who has written extensively on the close-up, argues that the power of the close-up is in the separation of the face from its actualisation in an individual person:

Ordinarily, the face of a human subject plays a role that is at once individuating, socializing and communicative; in the close-up, however, the face becomes an autonomous entity that tends to destroy this triple function: social roles are renounced, communication ceases, individuation is suspended.[7]

Audiences are relied upon to 'read' facial types, often stereotyped in terms of the beautiful being the 'good' and the disfigured being the criminal (or sometimes the super-heroic) as a result of their bitter experiences. In cinema and elsewhere, ordinary faces are what we identify with. Distortion of the face brings with it great power to disturb. Whether portrayed with a degree of sympathy towards the spirit trapped behind the monstrous visage or not, a monstrous cinematic face is usually used to indicate a monstrous identity and behaviour type.

In the earliest days of motion pictures, before the edges were smoothed out of the preferred film face, many faces were notably distorted or grotesque. Comic cinema was driven by the comically 'distorted' faces of actors like Ben Turpin (with his distinctive bushy moustache and crossed eyes), whilst figures like *Man of a Thousand Faces* (dir. Joseph Pevny, 1957) Lon Chaney dominated the horror genre. In keeping with the deployment of comic masks in ancient drama and 'satyr' plays, many of cinema's earliest films (that is, those made between the late 1890s and early 1900s) were driven by the gurning expressions of face contortionists and 'rubber-faced' entertainers.

The 1931 film version of Mary Shelley's *Frankenstein* (dir. James Whale) revealed a face never before seen in cinema. The image of Boris Karloff in his hideous monster mask, with bolts in his neck and wearing undersized clothes, has become part of worldwide popular culture. The shadowy profile in our first close-up look at the square head and expressionless face, with its sunken eyes and jagged surgical scar, is unforgettable. In 1939 Francis Pierce, Universal Pictures' chief make-up artist, decribed how he designed the mask:

I did not depend on imagination. In 1931, before I did a bit of designing, I spent three months of research in anatomy, surgery, medicine, criminal history, criminology, ancient and modern burial customs, and electrodynamics. My anatomical studies taught me that there are six ways a surgeon can cut the skull in order to take out or put in a brain. I figured that Frankenstein, who was a scientist but no practising surgeon, would take the simplest surgical way. He would cut the top of the

Jocelyn Wildenstein (p. 156)

Boris Karloff as **Frankenstein's monster**, from *Bride of Frankenstein*, 1935 (p. 157)

Tattooed face of Paul Morris, 1999 (p. 158)

LawickMüller, **Athena Velletri – Nina**, from the series *PERFECTLY SuperNATURAL*, 1999 (p. 159)

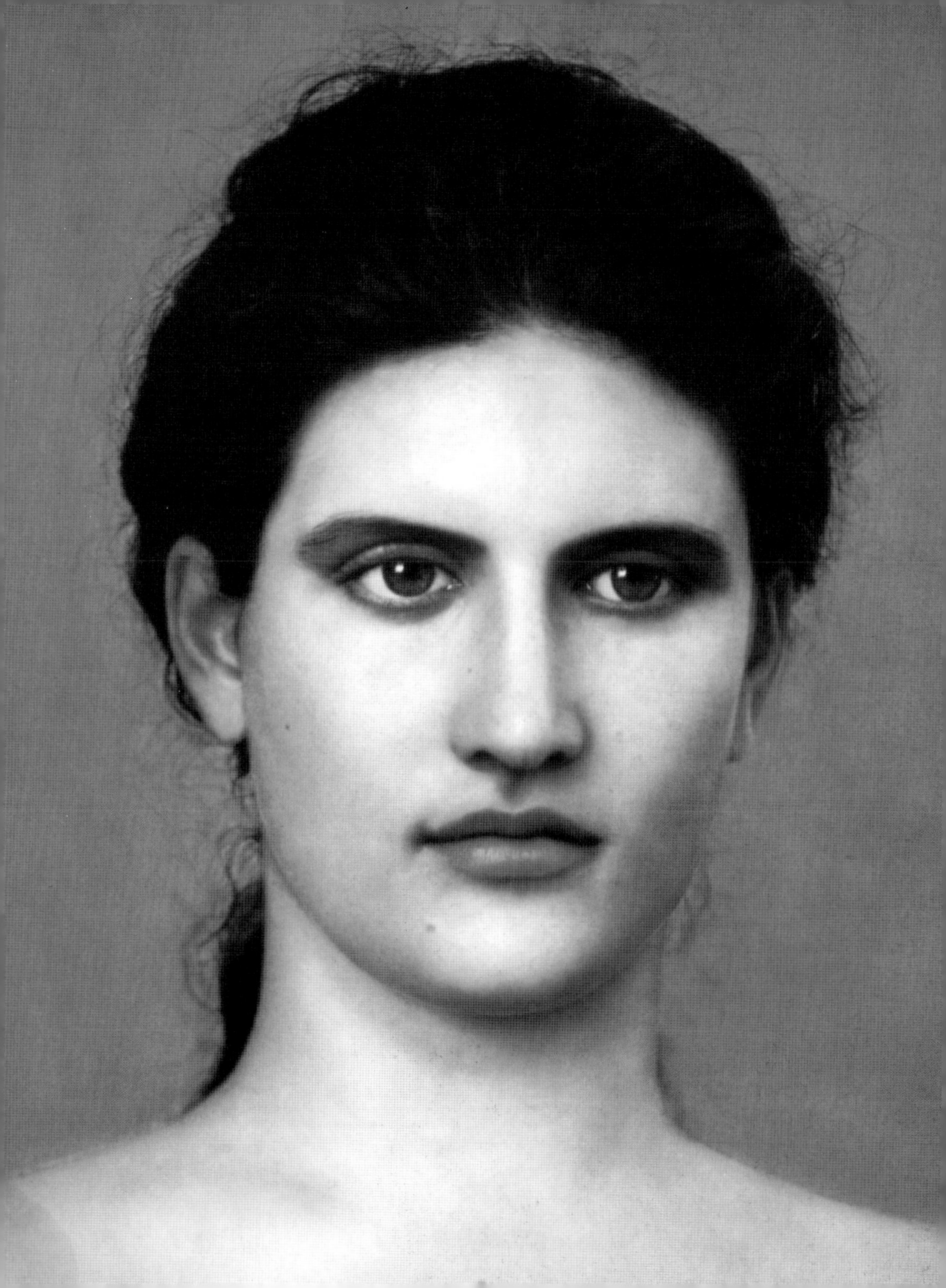

skull off straight across like a pot lid, hinge it, pop the brain in, and then clamp it on tight. That is the reason I decided to make the Monster's head square and flat like a shoe box and dig that big scar across his forehead with the metal clamps holding it together.[8]

The fear, loathing and horror inspired by the Frankenstein fable is partly due to anxiety about being destroyed by our own technologies and a movement beyond familiar reference points for 'normality' and a known world. Behind the fantasy of the machine made human, and a being simultaneously alive and dead, is both the fear of something that could destroy us and the exhilaration of the unknown. Like the killer robot played by Arnold Schwarzenegger in the *Terminator* films, the Frankenstein myth makes us question the stability of the known world, our sense of ourselves and of the laws of science and of human society as we know them. It is threatening psychologically, morally and socially.

David Lynch's 1980 film about the real-life Elephant Man also epitomises how we deal with the unacceptable and the inexplicable when it is expressed through the face and body. We find it hard to relate the Elephant Man's physical attributes (including his giant head, his crooked mouth and the lava-like eruptions on his skin) to our own physical appearance and so we would prefer to avert our gaze. The Elephant Man's physical condition also activates deep-seated medical anxieties and phobias and a fear of being outside society through no fault of one's own. His predicament symbolises the potential vulnerability and contingency of each individual life and a fear and unwillingness to transcend what is perceived as the norm.

dislocation

Grotesque or monstrous faces may scare or disturb us, but the 'face out of place' can be equally unsettling. In his essay 'The Uncanny', Sigmund Freud describes his experience of his own face as a double when he sees his own reflection in a swinging glass door. For a moment he doesn't realise it is his own face and feels hostility towards the man ahead of him. The sense the face produces is at once reassuringly familiar and disturbingly strange. It provokes a sense of horror because it resembles unfulfilled possibilities, a sort of parallel universe of what might have been. The same kind of suspension of the known world is conveyed in the following account of Harold Gillies preparing for facial surgery in 1917. The patient survived the operation, but died when a nurse applied Vaseline to the skin graft at too high a temperature and the graft necrosed. The post-operative photographs are unbearably painful to look at. The distortion of the face is horrific.

Major Gillies is about to operate. The patient's position is not quite suitable. He puts a yellow-gloved hand upon the patient's yellow shoulder and touches him. The effect on me is something like a shock. What was something like a man, seems of a sudden to be a figure stuffed with straw. The figure flops to one side, soulless, boneless.

'You understand what we are going to do?'

I shake my head.

Major Gillies points with his knife to the man's chest. There, faintly marked on the reddish yellow flesh, is the shape of a face. 'These spots here are the eyes, this is where the nose will be, and here you see the mouth which we shall give him.'

Good God, it searches me to the bone! That pencilled face on the man's chest, like a mask; and above that pencilled face on the chest, the old blasted and shattered face that a few days ago had the beauty and freshness of youth; why do surgeons speak of these things as a landscape gardener of his plans?[9]

Following both world wars, thousands of veterans had no choice but to undergo surgery and live with a post-operative mask-like face.

In the 21st century, as a result of these pioneering surgical experiments, it is now possible to seek to regain 'the beauty and freshness of youth' through elective cosmetic surgery. But go too far, and the post-operative 'aesthetic' face suggests a lack of humanness, and the made-up and masked face disturbs by its excessive perfection. Angela Carter's rewritten versions of wolf and beauty-and-the-beast lore in her collection of stories entitled *The Bloody Chamber* contain numerous examples of masked beasts and mask-like beastly faces:

Only from a distance would you think The Beast not much different from any other man, although he wears a mask with a man's face painted most beautifully on it. Oh, yes, a beautiful face; but one with too much formal symmetry of feature to be entirely human: one profile of his mask is the mirror image of the other, too perfect, uncanny.[10]

Another image of the too-perfect face is to be found in the figure of shop mannequins. In a number of twentieth-century surrealist works, department store mannequins were used to image the disturbing immobility of a beauty which has become a commodity. Man Ray and Salvador Dali used them as subjects, and the French novelist André Breton's work *Nadja* describes an illicit encounter with 'an adorable waxwork figure'.

In an article entitled 'Lasting expressions', mannequin collector Marsha Bentley Hale argues that the 'myriad' mannequin personalities that have been 'born' over the decades in shop window displays provide an insight into how we perceive ourselves today:

The variety of facial types, coupled with overall make-up and expressions, is almost as diversified as the shoppers that pass by these silent selling

Lester Gaba, **Cynthia**, New York, 1936 (p. 162)
Pierre Imans, **Mannequins**, 1930s (p. 163)

David Hopkinson and Shelley Wilson, **Beneath the mask**, (p. 164)
Cindy Sherman, **Untitled MP 316**, 1995 (p. 165)

pierre Imans
figures de Cire et Mannequins d'

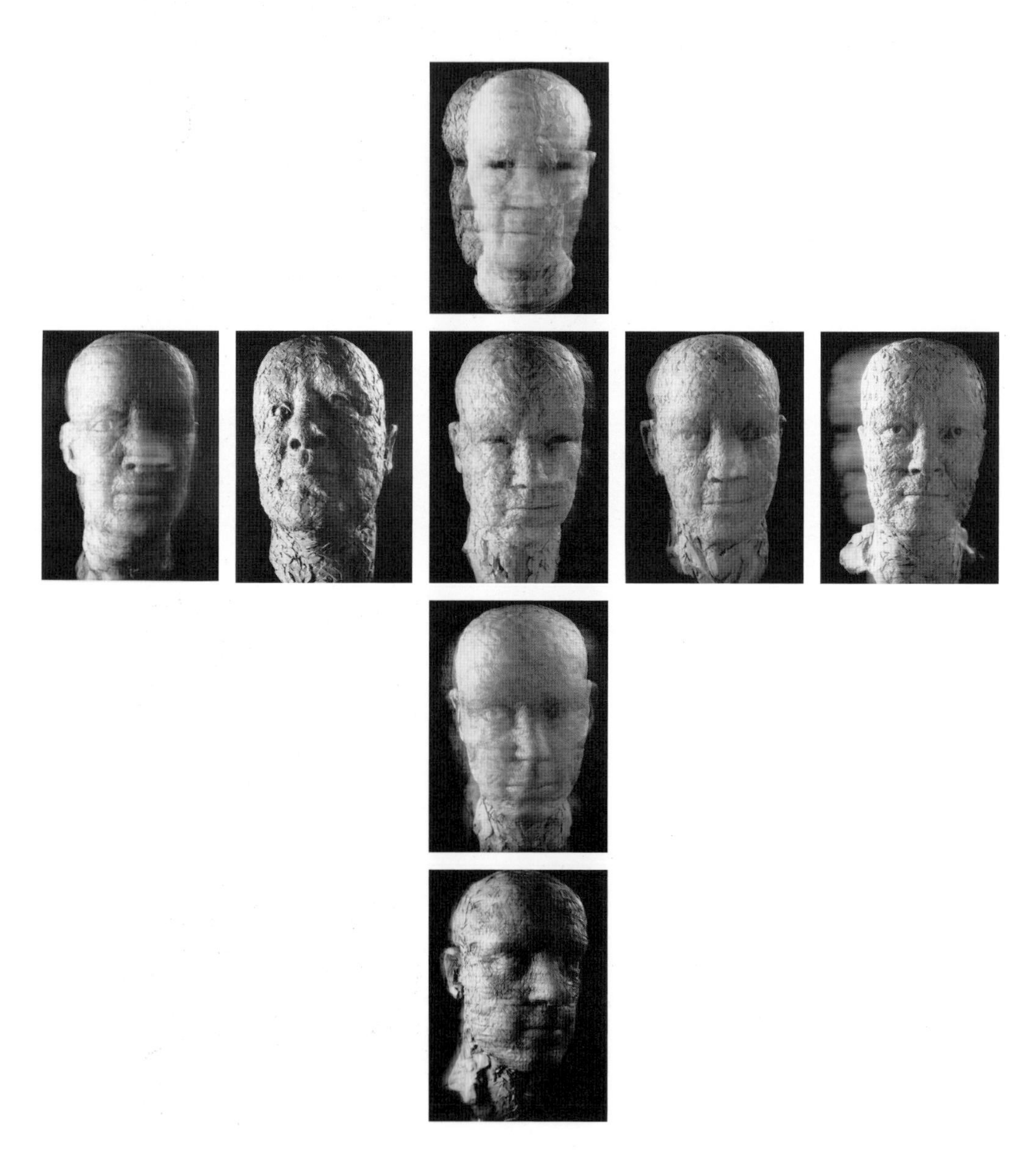

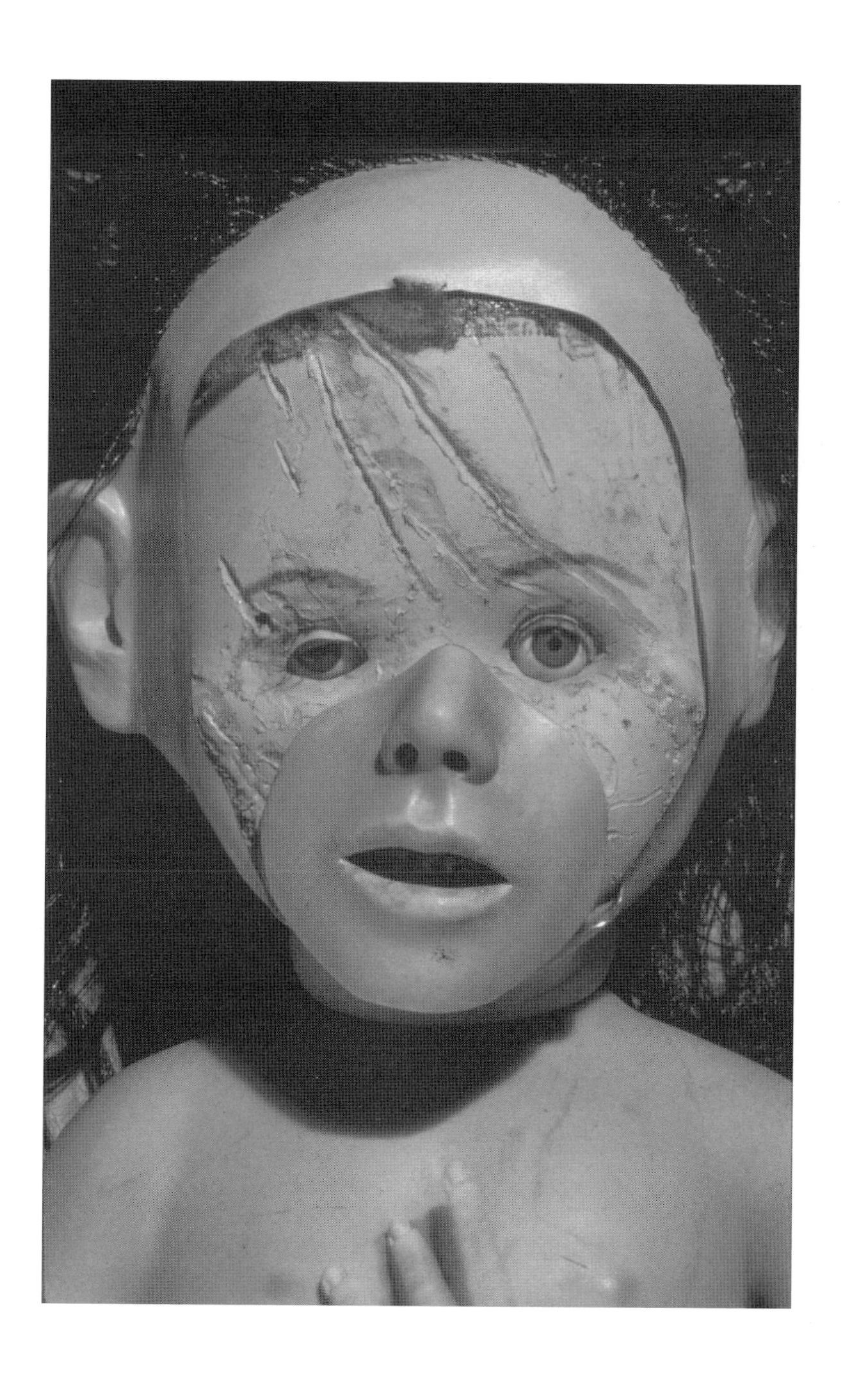

personae . . . Their expressions portrayed concepts of ideal images from the eras in which they were produced as well as telltale sociological imprints.[11]

First introduced at the 1894 Paris Exposition by German manufacturers, wax shop mannequins were both beautiful and extraordinarily realistic. Mannequins, male and female, appeared as props for photographs in *Vogue* and other magazines during the 1930s, where they are sometimes indistinguishable from the real-life models. The wax facilitated more lifelike expressions than either papier mâché (which had previously been employed) or fibreglass (used to fashion mannequins from the 1940s onwards). They were often implanted with real hair (cut from nuns), glass eyes and enamel teeth. The larger department stores had their own in-house facilities for remaking the faces of their mannequins, enabling them to update them quickly and easily.

Legend has it that mannequins like these were so lifelike that the renowned mannequin sculptor Lester Gaba supposedly even fell in love with one of his creations (or perhaps with the PR exercise that the relationship brought about!). He named her Cynthia and she accompanied him to all his fashionable New York appointments. Cynthia was seen in a box at the opera, riding in cars through the city and at night-clubs. Couturiers allegedly sent her clothes, and Cartier and Tiffany lent her jewellery. She eventually shattered when she slipped from a chair in a beauty salon.

Like the myth of the mannequin, there is something almost uncanny in the way many of Hollywood's re-cut faces are both reassuringly familiar and disturbingly foreign. While they attempt to hark back to a lost past – thus regaining the confidence and certainty of youthful good looks – they can become too 'plastic', horrifying manifestations of the impossible desire for youthful immortality.

dream

The dislocation of everyday reality and the scenarios of nightmare, dream and extreme distortion also frequently appear in film. Witness, for example, the scene in David Lynch's film *Lost Highway* when Fred Madison wakes from a nightmare, turns to face his wife in bed and sees a spectral grotesque face superimposed where hers should be. And in *The Silence of the Lambs*, Hannibal Lecter, typically shown in public scenes concealed behind a mask to protect the outside world from his cannibalistic bite, is able to escape by concealing himself beneath a severed – and, we imagine, chewed-off – dead face.

The face out of place, and the erasure of face as we know it, or facelessness, inspires horror. The plot of Georges Franju's 1959 film *Les yeux sans visage* is driven by desperation inspired by the improper destruction of a beautiful face; a brilliant scientist is guilt-stricken after having disfigured his daughter's face in a car crash. He has almost perfected the technique of grafting skin tissue and intends to use this science to rebuild his daughter's damaged face, but he needs a supply of donors to experiment on. His devoted secretary, Louise, lures women to his house and there, in a secret laboratory, the scientist attempts to remove their faces. Nearly half a century later, the same issue is discussed by writers on the internet chat-site of the London Science Museum Dana Centre's 'Face of the Future' exhibition: 'What if the face of your friend or someone in your family who had died in an accident was transplanted on to a new person?' posted one writer: 'You would see their face, but it wouldn't be them. It would be a new person wearing a familiar face. How creepy would that be?' 'Don't want to be gorgeous with cadaver tissue on my face,' writes another. 'Without my face, would I still be me?' writes a third.

Facelessness is not only a key factor within the domain of horror and monstrosity. Behind attempts to create 'believable' internet and interface avatars is the need to overcome resistance to disembodied interaction. (We use the words 'efface' and 'deface' as negative epithets.) Computer interfaces tend to adhere to an 'acceptable [humanoid] face' in order to facilitate the transaction between 'real' and 'virtual' worlds. Today the face as interface is as important in our dealings with machines as with other people. Mobile phone portrait messaging and online portrait albums are the 21st-century equivalents of the Victorian cartes de visite and daguerreotypes and of twentieth-century Kodak snapshots, paper photo albums and home videos. In this respect, despite virtual and global communication systems, the circulation of images of the face is as current as ever. Even in cyberspace, the visibility of the face remains crucial.

future face

Over the past century attempts to describe, analyse and even capture the face through art, surgery and scientific technologies have demonstrated its continued elusiveness. 'There's no art to find the mind's construction in the face,' as Shakespeare's Duncan says in *Macbeth*.

Looking backwards in time, rather than to future possibilities, offers only minimal assistance in understanding the face, or anticipating its future development, as much of the information about the evolution of the face is contested or unknown.

Where did faces come from? Early organisms had no faces. The evolution of bilateral symmetry led to animals with a front end and a back end; sense organs clustered at the front end, which would be the first part of the animal to come into contact with new environments. Neural structures and brains also evolved at the anterior, to handle the sensory information flooding in through the senses. Crucially, though, at some point the collection of sensory organs ceased to be just environmental detectors *taking in* information, but also became a way of *giving out* information. The face became a way not only of gaining a picture of the outside world but also of communicating with others, through identity and expressions: the assumption is that this is who I am and this is what I am thinking.

The face has been subject to multiple evolutionary pressures – on the development of the individual sense organs, for example, but also on the holistic face itself and its ability to convey information to others non-verbally.

From the moment they are born, babies appear to be hard-wired to look at faces and they very quickly learn to recognise them. Throughout life, adults are drawn to faces with the characteristics of babies and small children, perhaps evidence of an instinct to protect the young and vulnerable. Faces with worldwide popular appeal are often baby-like. Japanese 'manga' faces and Disney's Mickey Mouse are characteristic examples of this, and it is interesting to note that Disney recast the original face of Mickey Mouse to make it resemble that of a child. This is perhaps why we like puppies and other baby animals with wide-eyed, engaging faces as well.

It is also believed that, in order to enhance communication via the face, human evolution, in contrast with that of animals, involved the loss of facial hair on the upper and mobile or expressive part of their faces. Those people who are unable to see or to express themselves using their faces often express themselves and communicate through other parts of the human body. In an interview with the neuro-physiologist Jonathan Cole, Donna Williams, who is autistic, explained:

I could tell mood better from a foot than from a face. I could sense the slightest change in regular pace and intensity of movement of a foot. I could sense any asymmetry in rhythm that indicated erraticness and unpredictability. I could sense the expressive from the reactive. Facial expression, by comparison, was so overlaid with stored emotion, full of so many attempts to cover up or sway impression that the foot was much truer. I used sound in the same way, even breathing. Intonation aside, I could sense change in regular rhythm, pace, intensity and pitch. I think these things may be how animals make sense of people. Perhaps for them, as for me, this is a system that develops in the absence or delay of the development of interpretation.[12]

It is, however, generally agreed that the mechanical and expressive organs of the face increase our chances of physical survival, both through maximising the human ability to ingest and eat food and through the power of rapid communication. Our faces are capable of revealing some of our innermost feelings, but they are also expressive of something more. The extent to which the face consists of and reveals something other than merely physical and mechanical function is still subject to debate, however. For some, the face is the site of a person's soul, or spirituality, or life-force, and is reflected back to us when we look at someone else. This view is beautifully expressed in an unpublished lecture by the American philosopher Al Lingis:

How striking it is that most of the other animal species, when we encounter them – look at our eyes. When we come upon a rabbit or a grouse, they are frightened by our huge bodies. But it is not at our legs closest to the ground that they look, but our eyes . . . In the ocean, one day, I came upon a large octopus lurking in his cave. Peering into it, I saw his big eyes looking into mine. The body of an octopus, with its tentacles covered with suction cups which are also sense organs, with its absence of bones and its three hearts, is as remote from our bodies as one can imagine. Jacques Cousteau says we are closer to starfish than to octopuses. Yet the eyes of the octopus meets and understands our eyes. Understands our look, where there is nothing to grasp or conceive.[13]

In his 1982 book *Camera Lucida*, the critic Roland Barthes writes movingly of his search for a photograph of his dead mother. He explains how he could not find a photograph which expressed his image of her until he found a picture taken when she was five years old (that is, before he knew her): 'I studied the little girl and at last rediscovered my mother'.[14]

As Barthes' quest reveals, the compulsion to find and record the face of another is in part driven by the need for a visible token of that person (a kind of shorthand for them) in life and after their death. Faces may sustain our need for significance, continuity and permanence – we expect and usually want our children to look like us. But, for all that, they never entirely satisfy with respect to issues of personality and character. The face is fascinating and tantalising in equal measure. It simultaneously reveals and conceals.

For some, the face is conceived in God's image or is a trace of the divine in the human. For those who believe this, to see and to read God's face would be to understand the physiognomy of life itself, and it therefore follows that a number of religions forbid representations or images of the face of God. Meanwhile, writing of the face, Diane Arbus, one of the twentieth century's finest portrait photographers, noted: 'It's a secret about a secret. The more it tells you, the less you know.'

As to the future of the face, it would seem that we have reached a new stage of development in the 21st century. Perhaps the most intriguing question about the face today is whether it will continue to be shaped by natural forms of evolution, or whether we will ourselves determine its features, contours and future shapes. As we continue to manipulate, create and conceal, the new scientific and technological means available to us human beings may now supersede both nature and nurture.

medicine face

alf linney

medicine face

alf linney

Most of us are reasonably happy with our faces. Even so, have you ever thought what it would be like to change the appearance of your face? Not just temporarily, as with make-up, but permanently, as with surgery. For most people, this would be only a passing thought. For an increasing number of others, though, it's a must and can even become an obsession. In fact, more people than ever before are having their appearances changed by surgery. This can involve anything from a collagen injection to enlarge the lips to major surgery to change one's entire appearance.

Less fortunate are those people whose faces are disfigured because of an accident or a disease, or whose faces do not grow normally, due possibly to some inherited disorder. Although clever cosmetic disguise is sometimes used to minimise such facial disfigurement, more often the only choice is surgery.

Facial surgery considers of an area that other surgical procedures do not. The face is the most exposed part of the body and presents our most public image. Surgery must produce a face acceptable to its owner, which generally means acceptable to the rest of us. To specify what is an acceptable face is quite difficult. It's easy to say we know one when we see one, but identifying exactly what makes a face acceptable or not is more difficult. And if this is difficult, then saying why a face is beautiful is even more so, although there have been many attempts at establishing a framework for the aesthetics of the face.

All of this is very important when planning facial surgery for someone. People do like to know what they will look like after the surgery is done, and whether they will be deemed good-looking or not. Moreover, changing the appearance frequently changes personality, social behaviour and of course lifestyle as well. It is also worth reminding ourselves that although it is illegal to discriminate against any person on the grounds of ethnic origin, sex, sexual orientation, marital status or disability, it is still widely accepted that we judge people on the basis of their facial appearance. Indeed, it has recently been reported that UK employers ranked job applicants' skills and qualifications as less important than their appearance. The appearance of the face also affects its owner's psychological wellbeing, and this has a direct

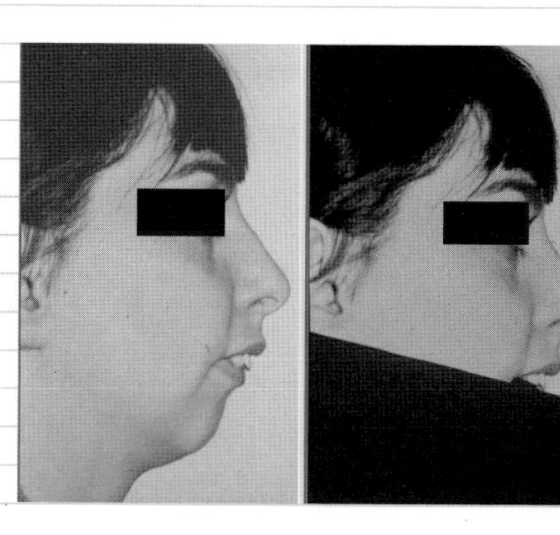
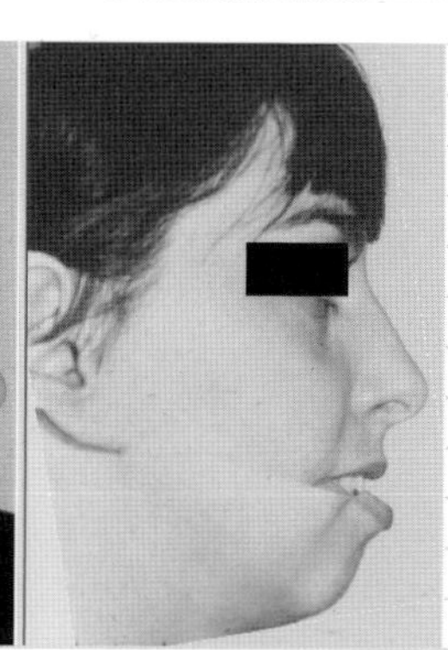

fig. 1
Using a photograph for planning surgery for the correction of facial anomalies

effect on the state of people's health. So for many, if not most people, the appearance of the face could fairly be described in broad terms as a health and social issue.

planning surgery

Over the years, surgeons and those who assist them have come up with a number of ideas and methods for planning surgery in a way that will give both surgeons and patients confidence in getting a good result, but there is still a long way to go.

Much of the early work centred on X-rays and photographs of the face. The X-rays were used to look at the skull bones that support the face and give it much of its shape. The photographs could be cut up and, subject to the constraints of what is surgically possible, put together again in a new way to show what the face might look like after surgery. Photographs are of course only two-dimensional, and the face has three dimensions. It's not very surprising, therefore, that this method often does not work very well, leaving patient and surgeon disappointed in the final result.

This system is illustrated in Figure 1, showing a young woman who has a very receding chin. The treatment is to slide the lower jaw forward, which is easily possible along the plane where the teeth meet, which can be found from the X-rays. The split and rearranged photograph gives some idea of how she would look in profile after the operation. There is, however, no way of showing the new full face appearance, although this will undoubtedly be changed by the operation.

In the 1970s the CT X-ray scanner was introduced. This allowed images of sections of the human head to be produced. Also during this period, some groups associated with facial surgery began to develop optical scanners that could record the shape of the face in three dimensions. These new technologies, along with the rapidly developing techniques of computer graphics, opened up the possibility of planning surgery in three dimensions rather than two, and also led to the possibility of simulating surgical procedures and predicting facial appearance after surgery. They have in recent years brought virtual reality as a guiding tool into the operating theatre itself.

fig. 2
Two facial scans of a child taken eight years apart

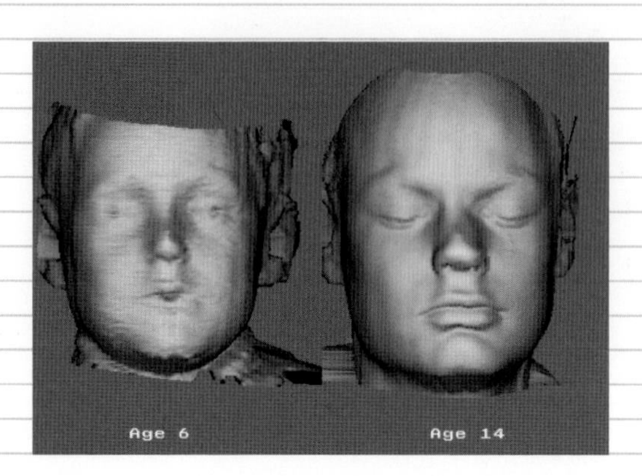

Some twenty-five years ago, the Medical Imaging Group in the Medical Physics Department at University College London started the development of a scanner for the recording of the three-dimensional surfaces of patients' faces. The purpose was three-fold: to be able to study the growth of the face for surgical or other treatment assessment, to be able to audit the effects of surgery or other treatment on facial appearance, and to assess the stability of the post-surgical face in long-term follow-up.

By rotating the subject under computer control, the whole of the facial surface is scanned, and more than 50,000 co-ordinate measurements are made on the facial surface with an accuracy of about half a millimetre. This takes about seven seconds, so that even quite small children can hold a facial expression long enough for a successful scan. The system is in regular clinical use and is installed in a number of maxillo-facial surgery centres. A face scan involves only a short exposure to a low-intensity beam of light, and is non-hazardous. This is most important as it means that the system can be used for long-term monitoring of a person's face many times over a long period, to provide information on facial growth and development.

Using computer graphics, the measurements are turned into a three-dimensional model of the patient's face. Just as with computer games, if we add a mouse or joystick we can interact with this model. We can rotate it to look at the face from any direction we wish, we can measure it, we can compare it with other faces and finally we can change its shape to explore what changes might produce the most pleasing or attractive shape.

Along with the scanner, the first computer programme to visualise the large number of measurements collected was developed in 1978. Both the scanner and the software have since then undergone continuous development, driven by an increasing number of clinical applications and raised expectations, as well as advances in video, electronic and computing technology. These changes are illustrated in long-term growth studies in which over the years we can see not only significant changes in the faces of subjects but also significant changes in image quality. Figure 2 illustrates such a study with two scans separated by approximately eight years.

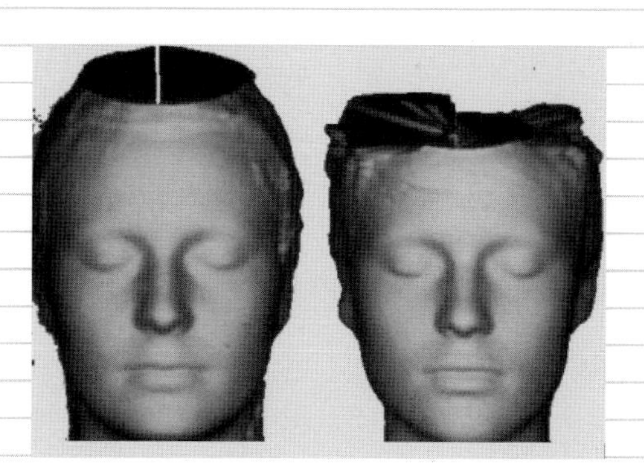

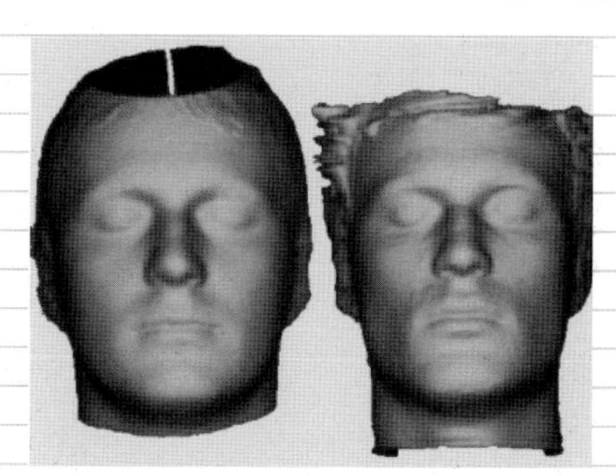

fig. 3
Average facial scans of a man and woman from the general population and the average scans for a group of photographic models

Because the faces are represented both by numbers and as images we are able to work with them mathematically. For example, we can calculate averages, and in this way work out the differences and similarities between two different groups of people. We have, for example, looked at the differences between the average faces of boys and girls at different ages. Once we have calculated our averages, we can then display them as images on the computer screen and appreciate visually the differences and similarities. People who are going to have an operation to change their face often ask if it is possible for them to look like a well-known photographic model or celebrity. This is presumably because the obviously beautiful are increasingly being promoted as the norm in television, film and video, not to mention the world of advertising. Because of this tendency, we have used the averaging facility to compare the average faces of photographic models with average faces for a group of people who are not photographic models.

Figure 3 shows the average faces of the man and woman from the general population and the averages for photographic models. We can see at once the special attributes that the faces of the models have which make them different from the average man and woman. It has been suggested that the models look more 'masculine' and 'feminine' than the others, and they may have been chosen as models for this very reason. Clearly anyone advertising a product would usually like the audience to be in no doubt as to the gender of the actors appearing in it. (There is admittedly a current trend for some fashion and fragrance advertisements to deliberately create uncertainty in the viewer as to whether the models are slightly masculine women or confusingly feminine men – but either way they are always glamorous in a predominantly feminine way.) Looking at these results, it is easy to see why craniofacial surgery is often requested by women wanting high cheekbones and by men wanting to strengthen the jawline and advance receding chins. Averaging is based on quite simple arithmetic. More advanced maths can be applied to the faces in order to give us another kind of information. The 3D surface of the face can be broken down into regions with similar shapes. For example, the cheeks and the chin are 'dome-shaped', under the lips we find a 'valley', and at the top of the nose is a 'saddle' shape. It is claimed that eight basic shapes can

fig. 4
The analysis of faces of family members:
father, mother, daughter and two sons

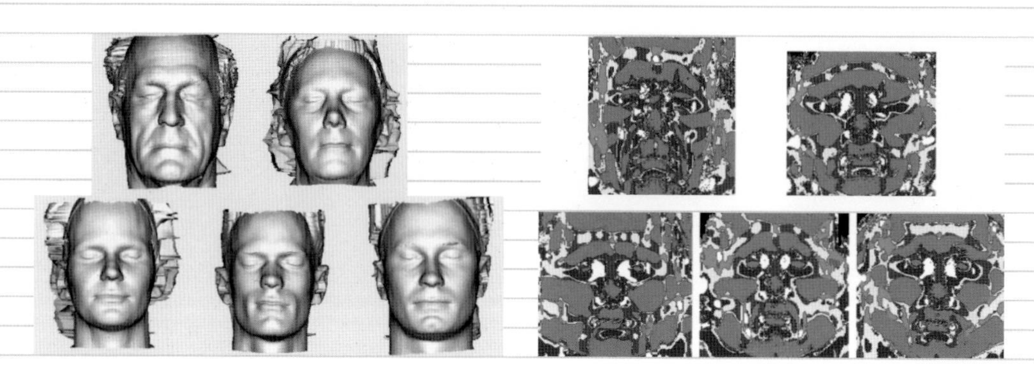

be used to build the surface of any face.

Another procedure involves 'painting' the regions of two faces in different shades to show what kind of shape they are. This is decided automatically by calculating by how much, and in what way, the faces curve at any point on their surfaces. You might reasonably be wondering why we would want to do this to a face! The idea is to come up with a description of a facial shape which doesn't depend on who is describing it. Earlier we mentioned just how difficult it is to describe a face, and indeed, if you ask half a dozen people to describe someone's face, you will get half a dozen different descriptions. The mathematical method, however, will always give us the same description. With these maths-based descriptions, we can go about trying to discover how certain aspects of our facial features are handed down from generation to generation – in other words, to discover something about how genes and facial shape are related. We know, because of the existence of identical twins, that genes largely control how we look. If we are able to learn enough about how genes control the shapes of our faces, it might even be possible one day to construct someone's face from a sample of DNA. Just think of the impact this could have on crime scene investigations!

Figure 4 shows an analysis of facial shape patterns for the investigation of inheritance. The mother and father are at the top and below them are a daughter and two sons. By looking closely at patterns on the faces it is quite easy to pick out bits of the sons' and daughter's faces that have been handed down from their mother and father. Using this kind of information, it is hoped that some inherited facial problems such as orofacial clefts, cleft palates and postaxial acrofacial dysostosis (Miller's syndrome) can be understood and possibly treated by influencing the genes. We are still a long way off doing this or reconstructing faces from DNA, however.

under the surface

The face is supported by the skull bones, whose shape is imposed on the softer tissues that could not hold their shape on their own. So in gross terms the shape of the face is very much determined by skull shape. Big changes in face shape can only be produced by altering the bones. Fortunately, we have

fig. 5
Reconstructing a 3D image from CT scans
fig. 6
Stages of surgical simulation to replace a tempero-mandibular joint

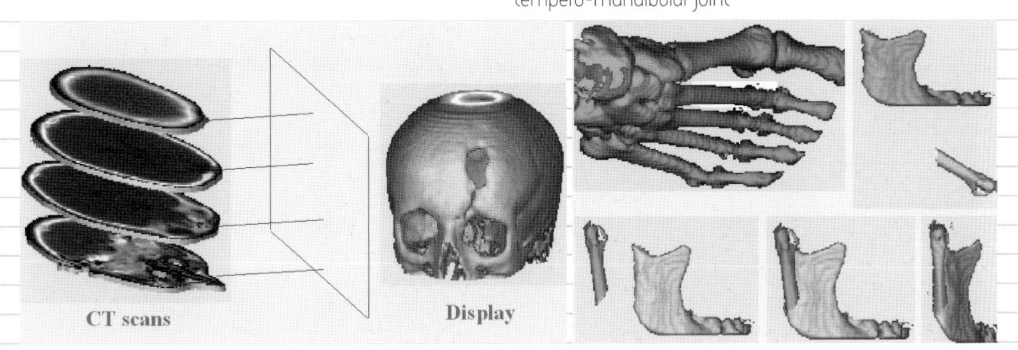

the means to get a good picture of these as well. Modern medical imaging systems designed to allow us to see inside the human body produce quite different pictures from the X-ray images with which we are all familiar. In the X-ray images, bones and organs are all piled on top of each other in a single image. The modern systems such as CT (computerised tomography) and MRI (magnetic resonance imaging) produce images of slices through the body in which we can see bits of the organs and bones quite separately.

Although these are considered to be of great value, and a radiologist can explore the body slice by slice to make a diagnosis, something extra is needed for the planning of facial surgery – we need to be able to see the anatomy intact as it is in real life. Developments in computer graphics techniques over the past twenty years have made this possible. The slices are basically stacked together to create the virtual 3D anatomy. We can then choose to look at the facial surface, the bones, veins and arteries, the nerves or even the spaces containing only air, like the windpipe. An image produced from 3D CT data is shown in Figure 5.

Further technical developments also allow the operator to interact with the image on the screen, making it possible to cut away various parts of the anatomy and in this way to simulate the surgery. The simulations are not very realistic at present, and there is no blood, but they do make it possible to see the most important things, such as the effects of cutting, reshaping and moving bones, and moreover to visualise just what it is possible to achieve.

One very positive result of these developments is that the patient can be very much better informed and so more confident in making the decision whether or not to go ahead with elective surgical procedures.

There are cases for which the use of an advanced planning system based on computer graphics simulation is absolutely essential. A good example of this would be the case of a woman who had asked for surgery to cure pain in the joints between the lower jaw and the skull. If you place your finger tip just around the point where the bottom of your ear joins on to your face and then open and close your mouth, you can feel this joint. This woman had previously

fig. 7
Result of the simulated surgery with jawbone in place

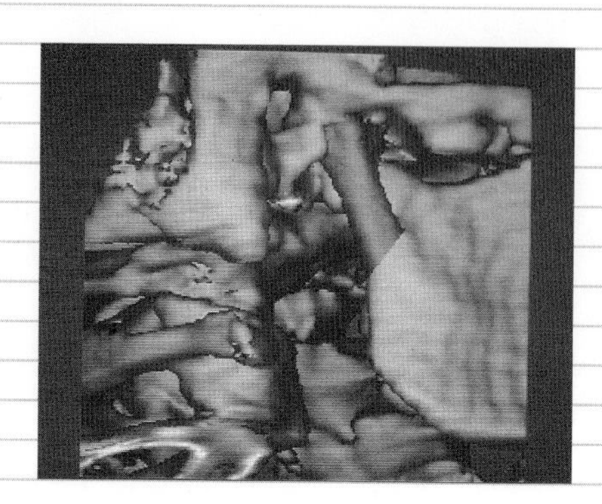

consulted a surgeon who had made a bad decision and had decided to cut off the 'knuckles' of the joints on either side of the face and put flexible pads into the joints. After only a short time, the pads sprang out of the joints, leaving the woman in even greater pain and unable to eat or speak properly. Her plight was so serious that it was decided to try a novel treatment, using knuckles from the feet bones to replace those removed from the jaw.

Such surgery needs careful planning. In the picture you can also see the woman's virtual feet. Using the patient's virtual skull and feet it is possible to simulate the surgery and to show whether it is likely to work or not. First the lower jaw is isolated from the skull, and the damaged joints are brought into full view. In Figure 6 the bone is extracted from the foot along with its joint and is trimmed to the size of the arm of the lower jaw. It is then merged with the jaw. Finally the jaw is placed back into the skull as shown in Figure 7, and it is even possible to check that the movement of the jaw is adequate. In this case the surgery was carried out according to plan and the patient was very pleased indeed, expressing her thanks in the national news media.

Although the simulation was vital in this case, it was limited to the bones. Experts, including Professor Nigel John at Bangor University, for example, are already working on future systems that will include the nerves, muscles and other tissues in the simulation.

Another surgical activity has been changed dramatically by the advances in technology described here. It happens that occasionally, for one reason or another, a person ends up with a hole in their cranium. An example is seen in Figure 8. The hole could be the result of an accident or of removal of a tumour. On the forehead, this appears as a facial disfigurement. One of the best ways of repairing such holes is to fix a thin titanium plate across them.

The use of titanium for repairing defects in the skull requires that the plate be formed before surgery. Before the use of computers, external impressions were taken through the soft tissue over the defect, and these were used to design plates. Unfortunately, this method does not generally produce a very good fit and the plates require manipulation with pliers and metal shears at surgery. This takes time, when time is very important; the less time the

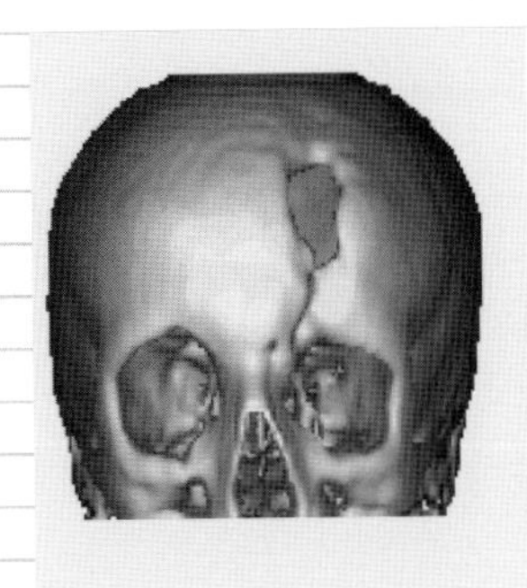

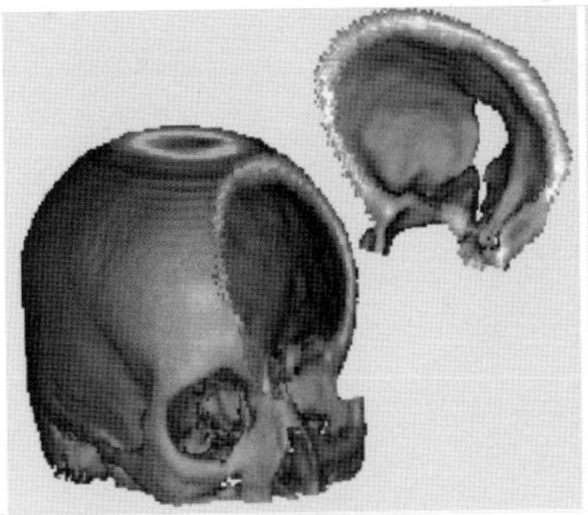

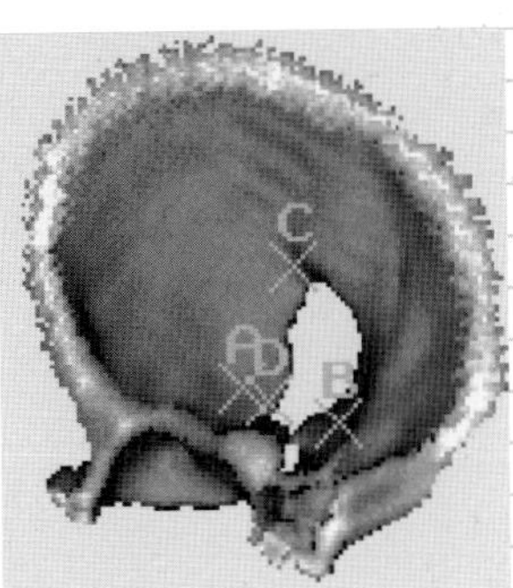

fig. 8
Example of details of a skull defect revealed by visualisation of CT scan data using computer graphics

patient stays under anaesthetic the better. Even after these adjustments, the plates don't always produce a good restoration of the shape or profile of the patient's head. By using a surgical planning facility based on 3D data and computer graphics, it is possible to greatly improve this situation.

Let's return to the example of a small skull defect, as shown in Figure 8. Using graphics for surgical simulation, we can cut open the skull (of the virtual patient) and inspect the nature of the damage from inside. The operator places marks on the bone surface in order to calculate the distances between them. The titanium plate is fixed to the skull using short titanium screws, and this capability to make measurements on the virtual patient is also used to give correct bone thicknesses around the edge of the skull defect, so that suitable lengths for the fixing screws can be decided before surgery begins.

The geometrical data can now be used to make physical models of the anatomy: a life-size model of the patient's skull and the defect. The model is used to cast a die in dental stone (a very hard kind of plaster), and the plate is pressure-formed against this. With this technique for manufacture, the time to perform surgery is frequently found to be reduced by up to one hour, and the fit and attachment are usually described as excellent. The technique can be applied successfully even when a very large part of the skull is missing. The restoration of the profile can be seen to be good and was described by the surgeon as excellent.

Representing the head by precise geometrical co-ordinates means that it is also very easy to produce a mirror image. This technique is used to restore a face when only one side is damaged or poorly formed. We can simply mirror the good side to produce a model for the damaged side. The repaired face will then have the expected symmetry we associate with attractive faces. This method is often used to repair cheekbones, eye sockets and jaws. Figure 9 shows a rapid prototype plastic model of a patient's jawbone that has been damaged on one side. The implant necessary to restore symmetry to the jawbone has been found by mirror-imaging the good side on to the defective side and computing the difference in 3D shape to make up the deficiency. This implant has also been made by rapid

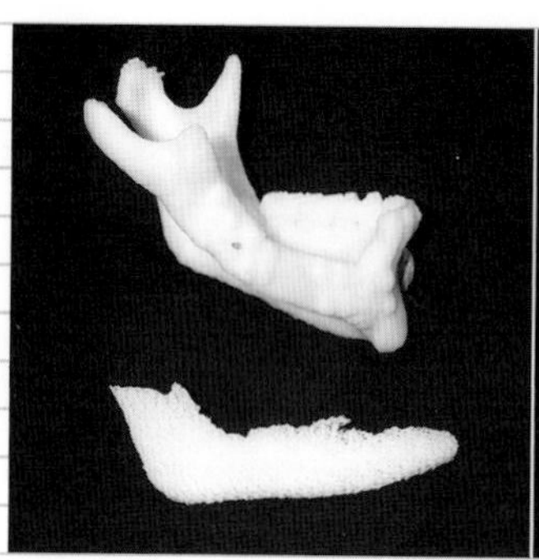

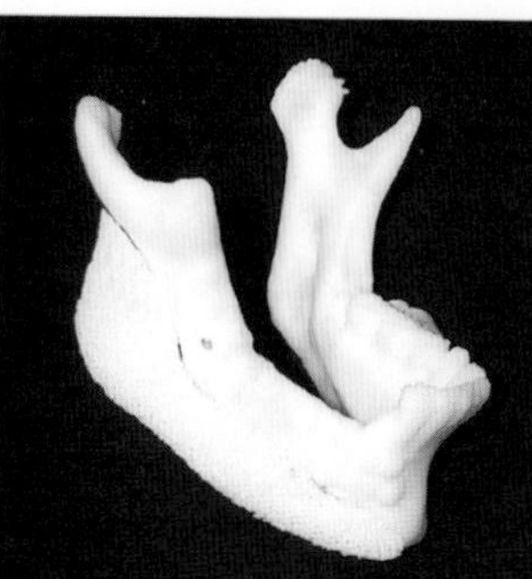

fig. 9
A rapid prototype plastic model of the jawbone of a patient which has been damaged on one side, along with the modelled implant design for its repair

prototyping and is used to shape a titanium mesh that will be used as the actual implant for the patient.

during surgery

We have so far looked at computer techniques for creating and using virtual patients for the study of facial growth and the planning of surgery, and for the manufacture of implants which have been designed to conform precisely to the individual patient's anatomy. This all happens before surgery and has been shown to have a beneficial impact on the results, as well as on the confidence of surgeons carrying out the operations.

With the speeds at which computers can produce images, it is now becoming possible to use the concept of virtual worlds and virtual patients to provide direct assistance to the surgeons during the operation in the operating theatre. This is possible by arranging for images of the real and virtual worlds to be mixed and presented as a single image. The product of this mixture is a highly informative environment known as 'augmented reality'. With this technique a surgeon can in effect be provided with 'X-ray vision'. Figure 10 illustrates this idea – a section of a virtual skull constructed from CT data has been mixed with a photographic image of the subject's head.

Surgeons have started to use this technique for surgery on the base of the skull and around the ear, an area where 'X-ray vision' is likely to be particularly helpful, as this part of the head has so many vital parts closely packed together that a very small surgical mistake can lead to deafness, facial paralysis or worse. It's also a good place to introduce the use of augmented reality, as most of the surgery is carried out using a binocular surgical microscope in which images of the real and virtual patient, can be conveniently mixed. Also, if we know where the microscope is, then we know where the surgeon is looking at the real patient, and can produce the appropriate matching image of the virtual patient. This allows the surgeon to see the underlying anatomy of the patient near the area of surgery, which cannot normally be seen with the naked eye.

This seems a fairly simple idea, so why isn't it widely used? There are still a lot of problems to

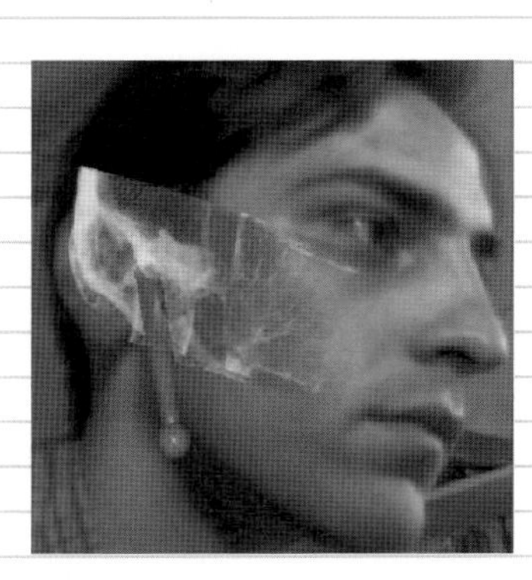

fig. 10
Mixing images of the real and virtual worlds: augmented reality for surgical guidance

solve. For one thing, we need to make sure that the real and virtual patient coincide in the position they appear to occupy in space. We do not want the surgeon to be looking at the real patient's left ear, but seeing the bones under the nose in the virtual patient image. In fact, to be useful, any system must be able to track the position of the real patient and to make sure that the virtual patient follows any changes in position very precisely and stays in what is called 'spatial registration' – in other words, matches the real patient's position, scale and perspective. In order to avoid cutting into vital structures such as nerves and blood vessels, the surgical tools being used by the surgeon also need to be tracked and shown in the augmented reality environment. Doing all of this is not easy, but there is a lot of work going into the investigation of virtual environments, and it will not be long before most of these problems are solved.

There are also a number of interesting spin-offs from technical developments in the application of computer graphics in surgery. When a skull is found in suspicious circumstances, the individual to whom the skull belonged has to be identified. To perform a reconstruction of the face, the skull is optically scanned. A scan is also taken of the face of an individual of a similar type (i.e. sex, estimated build and ethnic group) to that inferred from a visual inspection of the skull. The facial surface is then adapted to the skull using accepted tables of mean soft-tissue thicknesses. Relatives and acquaintances of missing persons are then invited to view the reconstructed facial image. The techniques developed for the reconstruction of faces over skulls for forensic purposes have also been used for archaeological purposes in the reconstruction of the face of a Viking for the Jorvik Viking Centre in York, and in a BBC series about our ancestors.

In spite of all these developments, it is still not possible accurately to predict the effect that surgery will have on facial appearance. We just don't know enough about how the soft tissues of the face respond to changes in the underlying skull bones. There is lot of individual variation in response to surgery, so the information will have to be collected on a large number of people undergoing surgery. Given the very personal nature of facial appearance, the task is not an easy one. Our control over future faces is thus still far from certain.

identikit face

vicki bruce

identikit face

vicki bruce

Like it or not, your face tells everyone else a great deal about who you are, and what you are thinking or feeling. Complete strangers can probably tell from your face if you are male or female, roughly what age you are, and something about your ethnic group. They may decide you look honest or shifty, intelligent or stupid. Your facial movements will tell them a lot about your momentary emotional and attentional states, and recent research suggests that individual faces may have characteristic patterns of movement that can provide additional cues to identity. How is information from the face used to help us identify people? What information allows us to classify faces by age or sex, to think we can guess their profession, or to decide that individuals look witty, or wise? How accurately can we identify people we have seen commit crimes? Can criminal identification become more accurate with aid from machines? Does the all-pervading CCTV camera make person identification trivial? These are some of the questions explored in this chapter.

what your face gives away

age

Our faces change dramatically as we mature and then age. Infant faces have small chins and noses, large eyes and brows – their features are relatively low in their faces. The lower face features in particular develop and grow through childhood, making the forehead and eyes relatively smaller in the adult face compared with the infant's. Further dramatic changes to the face are triggered through puberty, as male hormones make the mature male face display a range of sex-linked features (a large voice box, facial hair and hair follicles, enlarged nose and jaw, bushy eyebrows) which are missing in the female. The adult female face resembles the immature 'baby' face more than the adult male face does, which is perhaps one reason why the effects of ageing on facial appearance can appear more devastating to female attractiveness than to male. Long after sexual maturity has been reached, the effects of later ageing on the face become evident. During middle adult years, hair thins and greys, skin loses its elasticity and skin and flesh begin to sag. Facial skin itself shows wrinkles and other signs of age.

fig. 1
Which faces are male and which are female?
How do you decide?

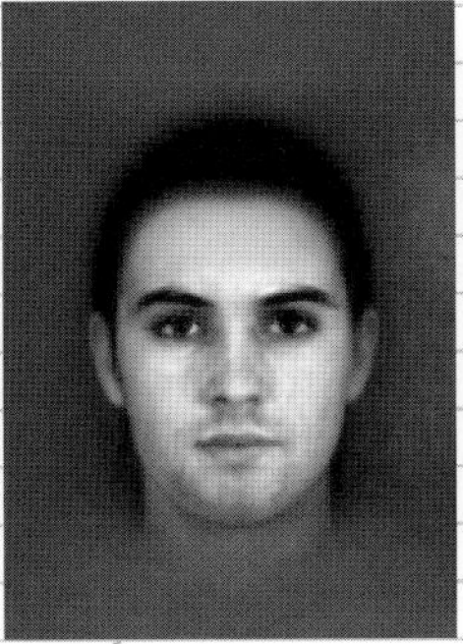
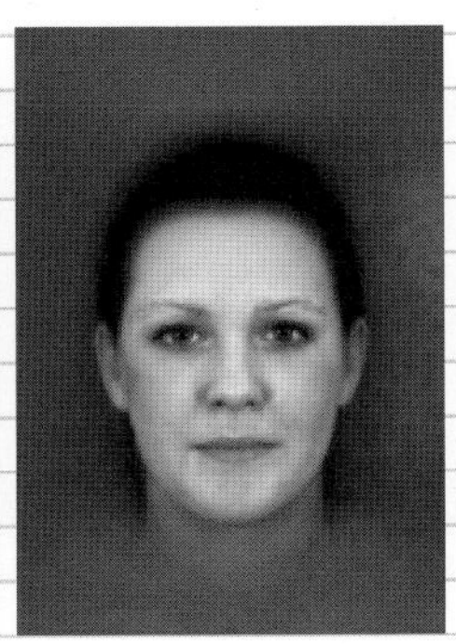
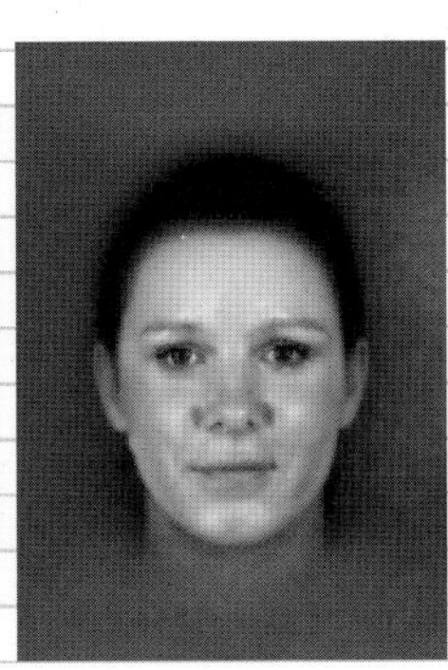

sex

Look at the face images shown in Figure 1. You will probably find it very easy to decide whether each is male or female, and research has shown that people are about 95 per cent correct in classifying images like these. How do you do it, though? There is not very much hairstyle to go on, nor do the male faces here sport facial hair (though their skin may appear a bit rougher than the women's).

Psychologists have examined what information allows us to make these categorisations so effortlessly. One method they have used involves making many measurements on a large set of female and male faces and seeing which measurements best differentiate between these categories. Psychologist Mike Burton and his colleagues[1] took this approach, and found that a set of sixteen measurements combined did the job of assigning faces to their correct gender groups as accurately as people. Some of the most informative measurements include those of the eyebrows and nose. (The male face has a more protuberant brow, with bushy eyebrows which are more extensive than the female's.)

Another approach is to show people faces in which some information has deliberately been concealed, and see which kinds of changes to faces create difficulties for observers trying to tell their sex. If eye and brow area is important, for example, then covering up this region should make sex judgement more difficult, and indeed it does, as research by a group including myself has shown.[2] Sex judgements made to faces shown with eyes closed and hair concealed with a swimming cap were 95 per cent accurate, but this dropped to just 86 per cent correct when the eye and eyebrow region was concealed with a mask. A third approach is to show people faces in which those cues which appear to be important are altered in some way. For example, as the distance between the eyebrows and eyes seems to be a very strong cue for distinguishing male from female faces, increasing this distance should make faces look more feminine and decreasing it should make faces look more masculine. Psychologist Ruth Campbell and her colleagues found – consistent with this idea – that faces looking down (which increases the apparent distance between brow and eye) were judged more feminine than those looking up.[3] Trish Le Gal and I followed up this work and looked to see how

fig. 2
The average female face made more feminine (a) and more masculine (b). The average male face made more feminine (c) and more masculi (d). Which image of each pair do you find more attractive?

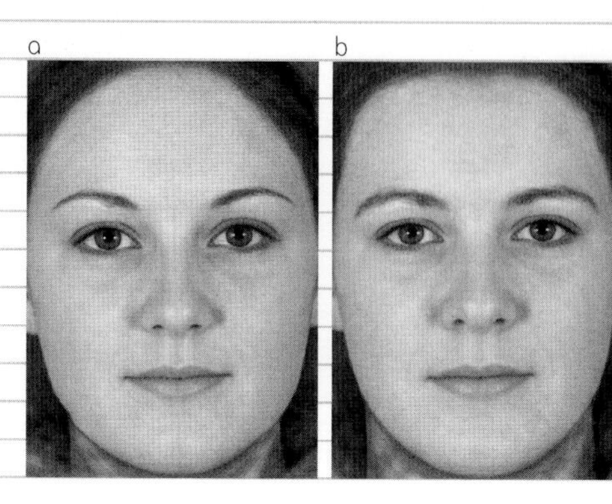

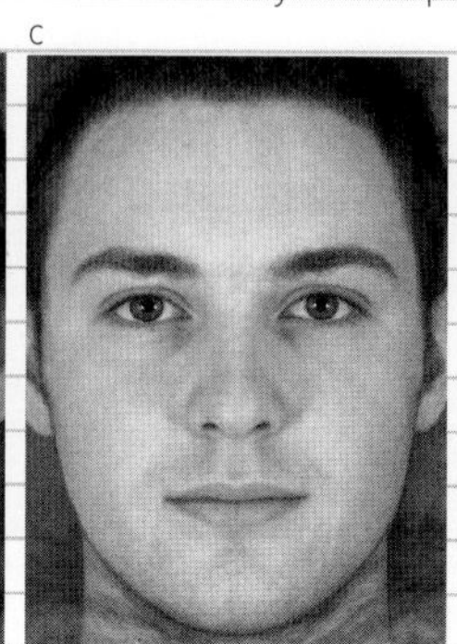

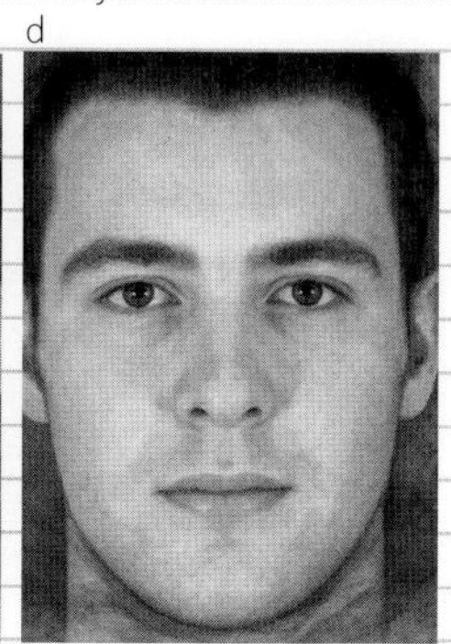

variations in expression of faces affected their apparent masculinity and femininity.[4] The facial expression of surprise increases the distance between eye and brows, that of anger decreases it. Surprised faces – whether male or female – should accordingly be rated more feminine and angry faces more masculine in appearance, and this is indeed what we found.

exaggeration and impersonation

Enhancing femininity and masculinity – to generate 'superwomen' and 'supermen' – can be done with computer graphics to produce amusing versions of celebrity faces made more or less masculine or feminine. Does enhancing femininity and masculinity make faces appear more attractive? Looking at the pictures shown in Figure 2, most people should find the femininised versions of the female faces more attractive. In systematic studies there is found to be a strong correlation between perceived femininity and attractiveness for female faces, and, moreover, strong cross-cultural agreement over which faces are more or less attractive. However, for male faces, the story is much less clear – and rather more interesting. Women seem to prefer males who are slightly more feminine in appearance (more 'pretty', perhaps), but this preference in turn varies with hormonal differences across the menstrual cycle. Psychologist Ian Penton-Voak and his colleagues have found that women at their most fertile appear to prefer more masculine faces compared with the faces they like at other phases of the cycle.[5]

Many women try to achieve such effects by using cosmetics, application of which can enhance the apparent femininity of faces. Eyebrow plucking, for example, increases the distance from eye to brow, and also reduces the amount of visible facial hair, which is a masculine trait, while make-up can be used to make the eyes look bigger and more feminine because they are more baby-like.

The artist Yasumasa Morimura uses make-up as well as costume and setting to recreate himself as cultural icons, both male and female. The examples of his dramatic impressions show that much important information is carried by areas beyond the face. We return to this later in the chapter.

In films such as *Tootsie* or *Some Like It Hot*, much of the comedy arises from the attempts

fig. 3
What did each of these men do for a living?
When did they live?

of male lead actors to impersonate a woman convincingly. It is interesting to consider whether Dustin Hoffman, Jack Lemmon or Tony Curtis makes the most convincing female, and to identify how they achieved these effects in relation to the distinctive features of male and female faces we have discussed above.

what do you do?

Look at the faces shown in Figure 3. Can you decide what the people do for a living and when they lived?

In fact, each image depicts the same profession – and each shows an average, rather than an individual face. The images show the average faces of the Newcastle United Football team in the 1920s, 1940s and 1990s. This demonstrates rather neatly how fashions in facial appearance change. In some eras it is fashionable for young men to wear their hair long, and sometimes facial hair is in fashion – sometimes not. What might appear to be the stereotypical face of one of the war poets in fact simply reflects what young men looked like in Edwardian times.

We are very prone to make all sorts of judgements about what people might be like, or do for a living, from their faces. People will be quick to judge if a face looks like a Labour or Conservative politician, or if it looks intelligent, honest and so forth. Moreover, there is a good degree of agreement between different people's judgements of the same faces. There is little or no validity to these judgements, however. We cannot tell if someone is a policeman or a criminal from their face, even though we may agree with each other that a particular face looks like a criminal's. If there is no validity in such judgements, where do our stereotyped views come from?

In Figure 3, the images were created by averaging together several different images of the same type (footballers from the same team), and the result was that some generic 'type' of face was revealed – though the types revealed say more about fashion than occupation. This use of facial 'compositing' to try to reveal the facial characteristics associated with a particular kind of person is a technique we can trace back to early work on photographic composites by the notorious Francis Galton, founder of the eugenics movement. Galton[6] thought that composites would help reveal the essential form of the 'criminal' face, and he was supplied with photographs of convicted criminals which he averaged together to seek support for his position. However, his results were disappointing – all his criminal composites revealed were the kinds of faces we might expect to see in any group of poorly fed people.

Galton's work on average faces, and on measurements on face profiles, also formed an early attempt to understand and measure faces scientifically in ways that were relevant to understanding and solving crimes.

who dunnit?

face composition

When a witness sees a crime she or he may be asked to try to build a picture of the criminal. Police artists may work with the witness to try to draw a sketch, but not all police forces have access to artists. For this reason, most forces will make use of one of many composite systems designed to help a witness create a picture of a remembered face. Early manual systems, such as Photo-fit and Identi-kit, worked by asking witnesses to select individual face features from collections and construct a composite by slotting these features into a template. Photo-fit used features cut out from photographs of faces, but differences in lighting and shading between the original photographs meant that the composite images could appear striped when the different features were placed next to each other. More recent, computer-based systems work on similar principles but allow features to be moved and blended in much more subtle ways – it is possible to build a much better likeness of a face in this way. However, even though composite systems can produce good likenesses when people are asked to copy a face directly, under realistic conditions of remembering faces, likenesses tend to be very poor. Figure 4 shows some composites of famous faces produced using different manual and computerised composite systems.

why are faces difficult to recall with composite systems?

To understand this requires that we learn a little more about how we normally perceive and remember faces. Composite systems such as Photo-fit treat faces as though they were a set of features – eyes, nose, mouth – that can be remembered and assembled independently. But much psychological research into face perception has shown that this is not the way we see faces – our impression of one part of the face is affected directly by other parts of the face.

A good way to illustrate this is to see what happens when two different halves of different famous faces are combined. This is what psychologist Andy Young and his colleagues did.[7] They showed that when the two half faces were aligned, volunteers found it extremely hard to decide to whom the top half of the face belonged. But when the halves were off-set, the task was relatively easy. This shows that the half face alone can be identified, but that when it is put together with the 'wrong' mouth and chin, a new identity emerges. This is one of many examples of the processing of the face as a configuration, rather than as a set of separate features.

It seems that we are 'expert' at perceiving the subtle variations in configuration that distinguish one face from another. If faces are turned upside-down, we lose the benefit

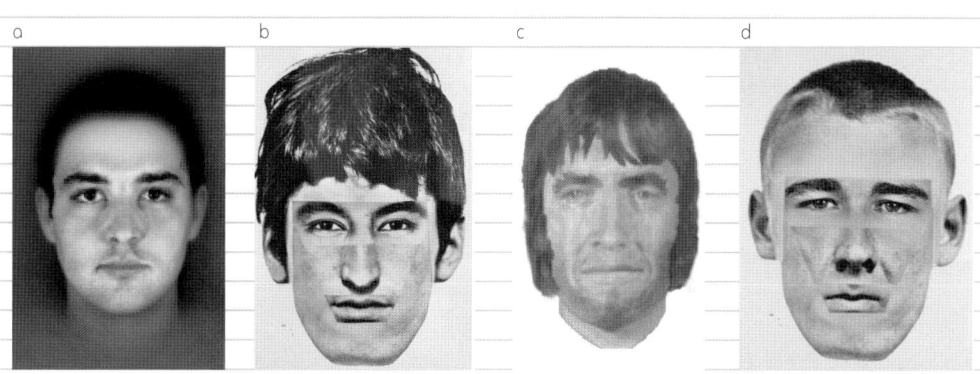

fig. 4
Which famous faces are these? Each shows a composite image produced with a different system

of this expertise with configuration, and process the faces in more piecemeal fashion. Interestingly, the composite effect described above disappears when faces are inverted.

Systems such as Photo-fit, and even the newer electronic Photo-fits, have two problems in this context. First, the witness is asked to describe and remember features separately – which is not how we remember faces – and second, the assembly of the face features into a composite may create the wrong impressions at a holistic level even if some of the correct features have been remembered.

Improving composite systems
The problems that witnesses experience using composite systems are now well documented. What can we do about this? Current research by Charlie Frowd, Hayley Ness, Peter Hancock and myself at the University of Stirling aims to improve composite systems to overcome some of these difficulties.

In one recent project, my colleagues and I investigated whether we could get better likenesses by combining the images created by different, independent witnesses, into a single composite by 'morphing'.[8] The technique of morphing allows us to make an average of two or more images of faces without the 'blurring' that occurs with photographic averages. Figure 5 on the next page shows four different composite images of the actor George Clooney, produced by four independent volunteers aiming to produce the best likeness they could of him. Alongside these four attempts is the 'morph', or average, of all four of these composites. Our research showed that morphed composites produced in this way were rated as good, in terms of likeness, as the best of the individual composites, and better than the 'average' individual. In identification experiments similar results were found. Why does averaging together the products of separate witness attempts create a better likeness? We reasoned that each individual witness will produce some correct and some mistaken features. Averaging their individual attempts together will tend to reinforce the correct features. Because there is no reason why different witnesses should make the same mistakes in their memories for faces, there will be more 'correct' information in common between the different witness attempts than incorrect information, and so the 'average' witness attempt will be a better

Answer: (a) and (d) are both Michael Owen: (a) using the newer electronic PRO-Fit system and (d) using the older Photofit system. (b) and (c) are both Noel Gallagher: (b) from Photofit and (c) from the electronic E-Fit system.

fig. 5
Four different composites of George Clooney from different 'witnesses'. At the far right is the result of averaging all four

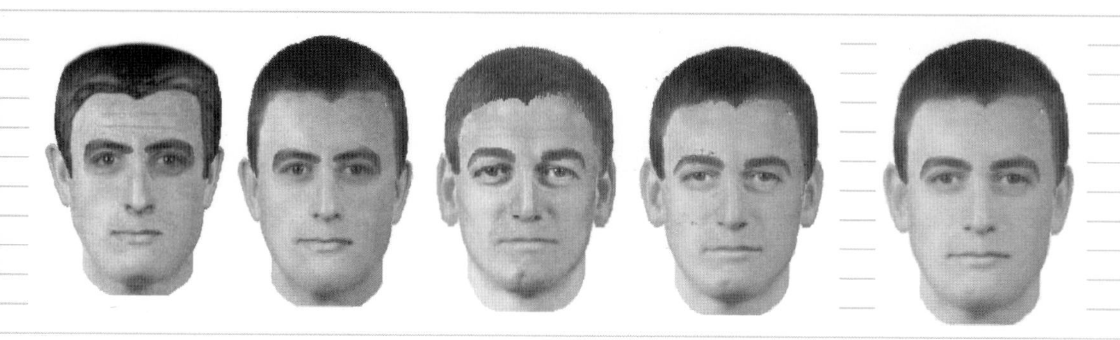

likeness than any individual attempt.

A second potential problem with the usual composite system is that the faces produced are very 'flat'. Working with a local company, we have been able to produce 'angled view' composites, and recently completed research has established that showing two composite viewpoints together is better than having just one. Figure 6 shows an example of how well a full-face and angled composite together can depict the likeness of an individual.

A third strand of research has explored a completely different method of building composites from witness memories. In developing 'Evo-fit', Charlie Frowd and Peter Hancock at the University of Stirling have suggested that witnesses might be better able to recognise than recall faces they have seen.[9] The system builds a screen of faces from which witnesses choose those most like the one they remember. From these choices, a new screen is generated using the technique of evolutionary computation – in which initial choices act as a first generation and then successive generations are 'bred' from these. Gradually the choices witnesses see converge on faces that more closely resemble their memories. One clever feature of this system is that the faces from which witnesses choose, while looking realistic, are not real faces but are synthesised from an underlying set of core dimensions derived from, but quite distinct from, real faces. This means that if Evo-fit were used in real criminal settings, there is no danger of any individual person being wrongly recognised from the screens of choices.

Evaluation of Evo-fit compared with existing composite systems indicates that it can perform quite well, though it has yet to be shown convincingly to surpass the conventional systems. However, ongoing work is extending Evo-fit to provide more variety in faces, and to show faces in angled views, and in full colour – see Figure 7.

CCTV images

The UK has more CCTV cameras per head of the population than any other country in the world. In theory, this should make it much easier to solve crimes, since in many cases cameras capture images of the people committing the crime. In practice, it can still prove difficult to identify people from such images.
CCTV images are most useful when their

fig. 6
Facial composites

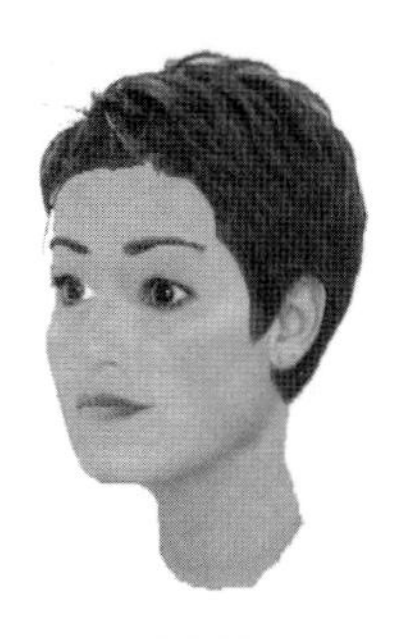
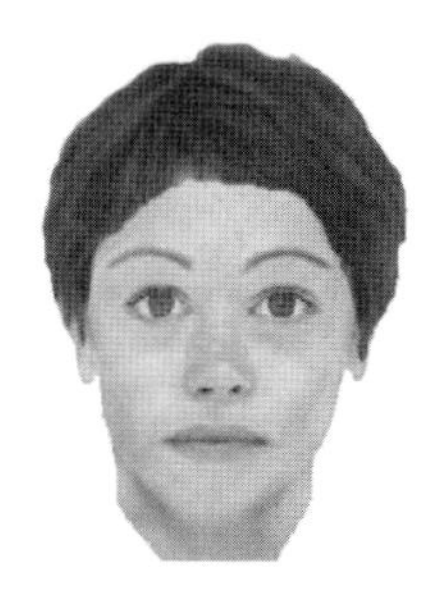

broadcast leads to spontaneous recognition by people who claim to know those shown. For example, the London nail-bomber was recognised in broadcast images by a former workmate, and this recognition triggered the further investigations which led him to be apprehended. Research has proved that we are extremely accurate at identifying highly familiar people from CCTV images even when the quality of the images is very poor. In one study, Mike Burton and his colleagues showed students video clips of familiar people captured on typical, low-quality security footage in which each face appeared extremely indistinct.[10] Students were highly accurate at identifying the people seen in these video clips, but only if their faces were visible. If faces were concealed, recognition was poor, even though cues from body shape, gait and clothing were still visible. If all that was visible was the face, recognition was surprisingly good. These findings suggest that camera quality need not be very good for CCTV images to be useful in helping identify who was present at a crime.

However, CCTV images are at their most dangerous when there is an attempt to use the picture recorded to 'prove guilt' by comparison with the image of the suspect. Other research by me and my colleagues has shown that two images of the same person can look very different, while two images of different people can look very similar.[11] This is because, when faces are unfamiliar, two images of the same face can look very different, and two images of different faces can look remarkably similar. When faces are unfamiliar, resemblance between images should be used to signal just that – resemblance – and not identity.

when face recognition fails

why are names so hard to remember?
When people experience difficulties in identifying familiar people – whether these are known personally, or are famous faces known via television and films – they often describe two kinds of problem. The first is that a face just seems familiar, but they cannot decide why. The second problem is that of retrieving names. There have been careful psychological investigations to back up these common impressions. For example, in one study, Andy Young and his colleagues asked 22 volunteers to record systematically in

fig. 7
A screen version of the three-quarter view version of 'Evo-Fit', a new technique for building composite images

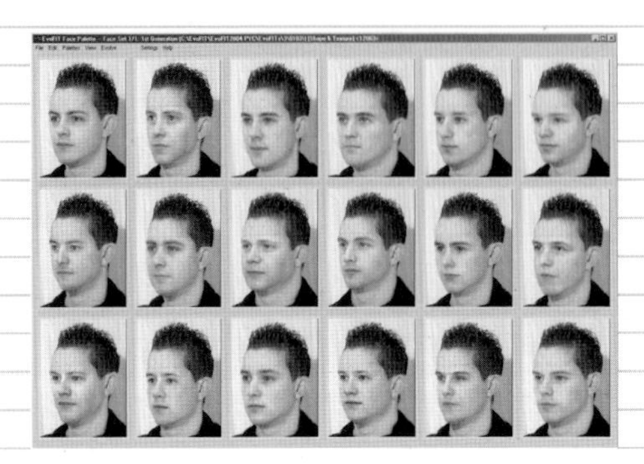

diaries all the difficulties they encountered in their daily lives when trying to identify people they knew.[12] Participants frequently had problems in knowing why faces seemed familiar, or in remembering names. Some incidents led to socially embarrassing episodes. Here are some of the records noted by their volunteers:

'I stopped a passer-by to ask directions. She looked familiar and then spoke as if she knew me. When she looked as if she knew me I pretended to recognise her too, but didn't ask her name.'

'I had gone to meet him, but I couldn't remember his name! I got very annoyed with myself, but didn't ask because it had been mentioned a few minutes before.'

So it is common to find a face familiar, but not know why, and it is common to remember where a face is familiar from, but forget the name. Some kinds of difficulty, however, are never observed. In particular, we never find people who can remember the name but not other things about the person. We never say, 'That's John Lennon – who is John Lennon?' This observation, that names are the hardest thing about personal identity to retrieve, has led to some ingenious experiments exploring whether it is something in the nature of names that makes them hard to recall.

It doesn't seem to be word difficulties that create the problem, however. For example, experiments have shown that people find it much easier to learn that a new face belongs to 'a baker' or 'a miller' than to learn that the face belongs to 'Mrs Baker' or 'Mr Miller'. The relative difficulty of names is found even when there is no need to speak the name out loud. For example, volunteers can be asked to press one button to faces called 'John' and a different button to faces called 'James'. Such identifications are made more slowly than those based on occupation or nationality. It is in the relationship between the person and something that is a name that the problem lies. Many theories of name retrieval have been proposed and none is entirely satisfactory.

A further common problem that we all experience is the difficulty we have in recognising faces of people from races other than our own. Caucasian people in Europe, for example, report that Chinese or Japanese faces look very similar to each other, and we find them hard to identify.

There are a number of possible explanations for such an effect, and experiments can help us to determine which explanation is correct. One possibility is that Japanese faces are visually more similar to each other than European faces. If that were so, then Japanese people should also find European faces easier to recognise than Japanese ones. In fact, Japanese people find faces of their own race easier to recognise than faces of other races, so relative similarity of the faces cannot be the whole explanation.

A second possibility is that our difficulty in recognising faces of other races is due to our relative lack of experience with these faces. If this were correct, then people who have been exposed to faces from many races in their daily lives should do better at other-race faces than people who have lived in a more homogenous community. There is some evidence for this suggestion. For example, psychologists Paul Chiroro and Tim Valentine compared recognition of white and black faces by black African participants with high and low exposure to white faces, and by white participants with high and low exposure to black faces.[13] Although each race found faces of the other race more difficult than their own to recognise, the difficulty was reduced for participants with higher familiarity with the other-race faces.

The cross-race effect is an important one in eyewitness testimony, since witnesses will often be required to identify the face of someone from another racial group. We should be particularly cautious about accepting evidence from a witness claiming to identify someone from an unfamiliar racial group after only a brief view of them.

brain damage and face recognition

While most people manage to recognise faces most of the time, some rare neurological deficits can lead to extreme difficulties or abnormalities in face processing. Sometimes, as a result of an accident or a stroke, people lose the ability to recognise faces that they know – even faces of friends and family. Their remaining ability to recognise people from their clothes, voices or other cues shows that they haven't lost their memories of these familiar people – it is the face recognition trigger which has been lost. This condition, known as 'prosopagnosia', is extremely distressing to the individual patients. But careful investigation of which abilities are impaired, and which spared, in such patients can help scientists to understand how face perception proceeds in undamaged brains, and may also help suggest compensating strategies for the patients themselves. For example, many prosopagnosic patients are able to interpret facial expressions completely normally. This is one important piece of evidence that facial expressions are analysed in a separate part of the brain from that which does the job of face recognition. They can also recognise people from their voices or clothing. Supplementing face recognition with other information about personal identity may be important in other contexts too.

more than just a pretty face

When face recognition fails, can other information be used to aid person identification? Recent research by psychologist Karen Lander and me has suggested that the characteristic way that an individual face moves may also help us to recognise when the face is seen under difficult conditions. This research took as its starting point an observation from Alan Johnston and his colleagues at University College London that faces seen in photographic negative – which are extremely hard to identify – become easier to recognise if they are shown moving. Now that observation alone could merely indicate that the additional information in a moving clip (which shows a variety of poses and expressions as a face moves and speaks) can compensate for the loss of information in the negative image. But Karen Lander went further to demonstrate convincingly that it was the movement per se that was important. She showed videos of famous faces which were degraded to make them difficult to recognise, and compared how well they could be identified from the original videos against ones that had been slowed down or had the frames rearranged so that the movement was incoherent. If a moving image is easier to recognise because of extra information, then it should make no difference if the speed or order of the information is changed. Karen found that showing the videos in their original speed and form gave a significant benefit compared with slowed down or rearranged videos, suggesting that we may remember something about the characteristic movement of individual faces that helps us to identify them.

Indeed, this may make some sense of the compelling impressions given by impersonators of famous faces even when their resemblance to the target is fairly light. Impersonators who do a great job of the voice, mannerisms and gestures may also be helping to compensate for their lack of facial resemblance at the level of face features alone.

automatic person recognition

The use of information additional to the face pattern alone is also likely to be necessary if person identification is to be done reliably by computer. Although there has been some much-publicised use of automatic face recognition linked to video cameras, and although computers have the advantage of being quick and tireless at scanning and comparing images, current computer algorithms for face recognition are no more reliable than humans at comparing face images. So current attempts to build more reliable biometric data into passports, for example, will eventually need to supplement the use of face images with additional information.

From mid-2005, UK passports will include a digitised face image linked to a chip contained within the passport. Some time later, such passport chips will also contain other information to identify the person uniquely. This other information may be from fingerprints, but iris patterns are also possible. The advantage of iris patterns compared with face patterns is that irises have 249 different dimensions of variation, compared with only about twenty for faces.[14] In tests of iris-based recognition, there have been no cases of confusion between different iris patterns, compared with frequent confusions when face patterns are used. This leads to the suggestion that a person's iris could be scanned and stored on the passport chip, and a simple camera shot could be used to verify identity at an airport. Fingerprints are also attractive in terms of their uniqueness, but likely to prove much less socially acceptable, since fingerprinting is associated with criminal investigations. Iris scanning is currently being used experimentally at a number of airports in Canada, the UK and the Netherlands, and there are plans to expand these numbers. Iris recognition is also likely to be the key to the use of automatic verification of identities in other settings, such as cash dispensers.

is big brother watching you?

What are the social implications of such technological developments? If your iris is linked to your passport and scanned at each port of call, will your every movement be recorded centrally? How will society reconcile the possibilities of linked biometric information and personal movements with the very strict limits placed on storage of personal information through the Data Protection Act? Will there eventually be moves to link your facial appearance with your DNA? If automatic recognition of people via street CCTV cameras were improved, what are the implications of that for our personal privacy? Can any of these automated systems be useful for the prosopagnosic patient, stranded in a world of strangers?

conclusion

This chapter has given a very rapid introduction to some of the research conducted by psychologists that has helped us to understand how the human brain deciphers the pattern of the human face. Every time one human looks at another, they are able rapidly and automatically to use subtle cues from the visual pattern of the face to deduce whether it is male or female, old or young, attractive or unattractive, familiar or unfamiliar. We are some way off having machines that can make these judgements as well as people can, though human perception is itself fallible. When we consider how many faces we look at, and how similar they all are one to another, perhaps it is remarkable that our failures are so infrequent, and that our face-to-face interactions and deductions generally proceed so smoothly.

Elastick skin man (p. 196)

Anthony Aziz and Sammy Cucher, **Dystopia, Maria**, 1994 (p. 197)

JAMES MORRIS
ELASTICK SKIN MAN.

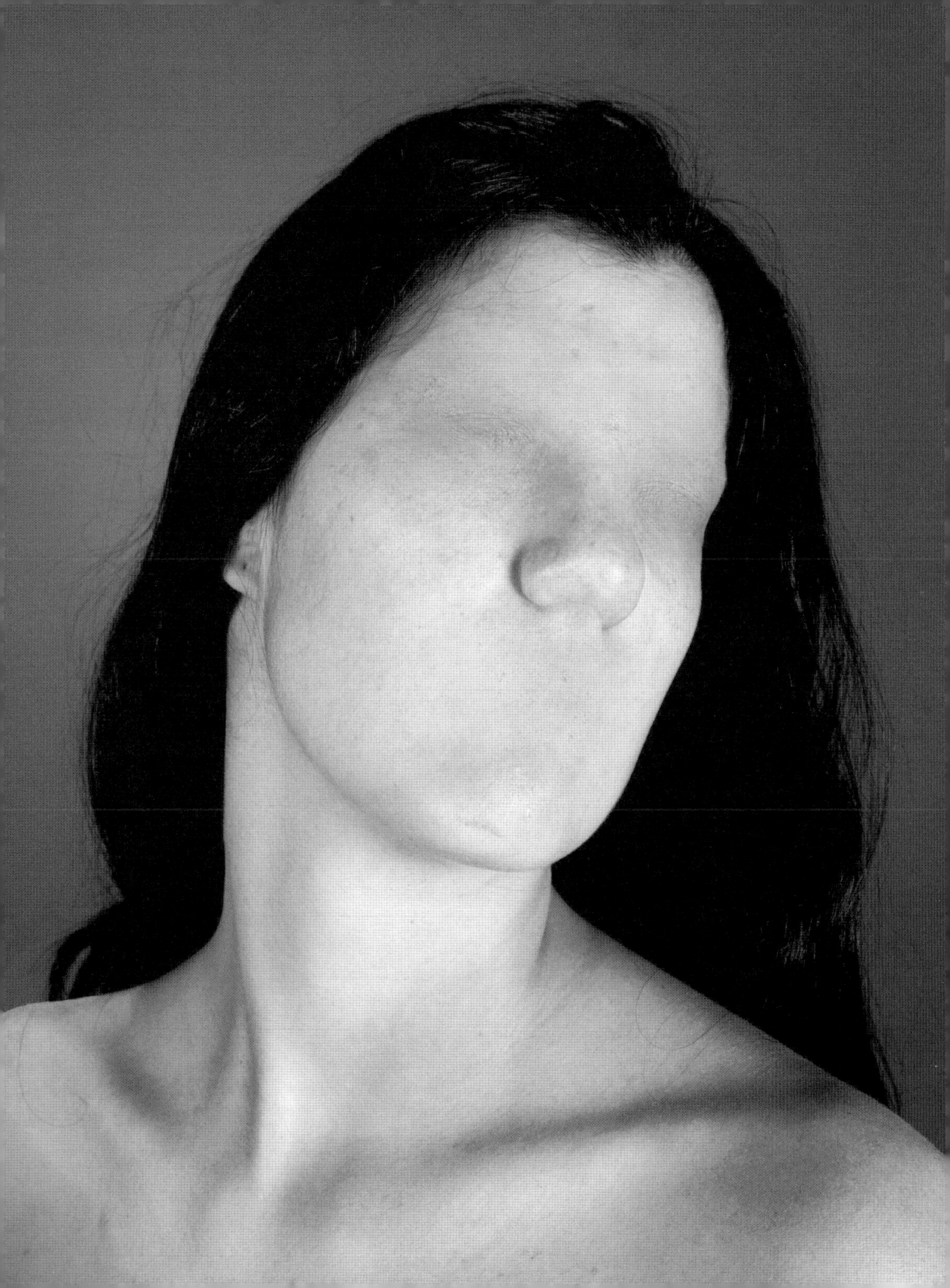

conclusion

sandra kemp

In 1974, in his book *The Human Face*, the psychologist John Liggett announced that:

> The time is not too distant when the questions linking face and character will be written with certainty – when we shall be able to relate truly accurate descriptions of facial form, textures and movements with genuinely objective movements and analyses of personality. We must await the development of an entirely new technology of facial description and analysis, which matches in power and precision the science of personality. Then we will be able to pronounce with certainty on the relationship of face and character. Then we shall know whether and to what degree certain aspects of facial structure and movement are serviceable guides to personality and experience.[1]

Knowledge of the face has been building for centuries. This book has explored aspects of that development in different disciplines and through the integration of the arts and sciences. Today we have increasingly sophisticated ways of understanding, reconstructing and representing the face. Scientific progress has led to the development of complex biometric systems to measure, classify and record the face. American scientists are about to perform the first human face transplant and applications to clone human beings look likely to succeed in the near future. Through technological progress, particularly in digital technologies, facial images and representations in film and in artworks have become more and more lifelike and autonomous. In contrast with their equivalents in analogue, these digital images are now able to exist, develop and replicate without a link back to the living. But, despite all this, in some profound way the face and its relationship to our innermost selves remain unknowable. 'So easy to look at, so hard to define', as Bob Dylan's lyric says.

Perhaps the clue to this continuing mystery has something to do with the fact that faces have evolved in tandem with our ability to read minds, and we still have so little understanding of the workings of consciousness. Recent scientific research has claimed that mentalising, or imagining how another person's mind works, is what distinguishes human beings from all other species of animal, and that it is an integral part of the complex rituals and power structures of social exchange.[2] It is significant how much we can convey through our faces, and the human face is a powerful tool in an increasingly socialised world. There would, after all, be no point in conveying fear, anger, happiness and so on through our facial expressions if no one around us was interested or understood them. We could communicate those emotions just as effectively by shouting, laughing or waving our arms around. Elsewhere in the animal kingdom, vocal utterance or gesture appears to do the trick. It is possible that the full meaning of the face may continue to evade us until we have unravelled the workings of human consciousness and have a comprehensive theory of mind.

Likewise, whether in traditional art forms or in the newer ones which are exploiting all the developments of technology, the human subject as an image remains central – and problematic. Many of the questions about whether the portrait can or does reveal a complete inner being or outer likeness, which have been debated in art, science and philosophy from Aristotle onwards, are still unanswered. Reflection on the future of the face, and on what the consequences of reaching a greater understanding of it would be, is itself Janus-faced. As the narrator of Kobo Abe's novel *The Face of Another* says:

Isn't it a preconception derived from habit to suppose that the soul and the heart are in the same category and can be negotiated only through the face. Isn't it common to find a single book, or poem, or record that communicates with the heart far more profoundly than a hundred years of scanning faces. If a face were indispensable, a blind man couldn't know human characteristics, could he? . . .

'But I'm asking you, every time I daydream about people on other planets I wonder why in heaven I always start with speculating on what they look like.'[3]

Portraiture as a genre continues, despite repeated assertions of its extinction. So rooted is the face in our consciousness that, at the moment, it seems likely that we will go on wanting to change, capture and reproduce it.

A portrait is a representation of a face with something wrong about the mouth,' an old definition goes. There is no doubt that artists and scientists will go on trying to adjust the face in life and in art and to get it 'right' according to cultural conventions. Meanwhile, however, there is no doubt whatsoever about the enduring power of the face and its extraordinary complexity.

image narratives

I have chosen the visual material in my four chapters to complement the exploration of the relations between face and identity in the written text. The captions below also relate to the three pivotal questions. What is a face? How do we portray and interpret faces? What will faces look like in the future?

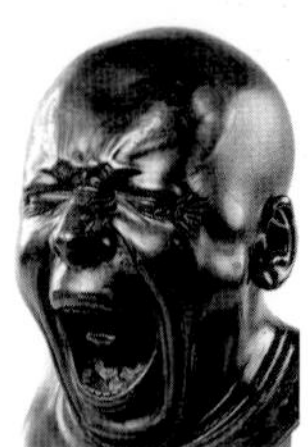

p. 8
Franz Xavier Messerschmidt,
The Yawner, detail, circa 1777–83.
The sculptor Franz Xavier Messerschmidt had a lifelong interest in conveying the expression of emotion in the human face.
After a psychiatric illness, possibly schizophrenia, in the 1770s, Messerschmidt began his series of 'character heads', such as *The Yawner*, which play with the extreme plasticity of the human face, distorting it into a gallery of grimaces. The sculptor painstakingly observed the minutest movement of the facial muscles with no thought for aestheticisation. Messerschmidt's work stands outside the formal conventions of Baroque portraiture. His contorted, near caricatured, faces are a playful exploration of human emotion and seem very contemporary, perhaps even foreshadowing Bruce Nauman's video installations.

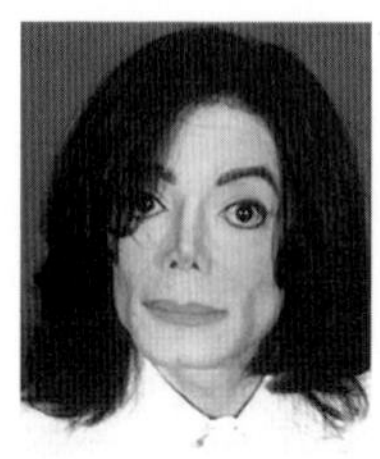

p. 10
Michael Jackson,
police photograph, 2003
In November 2003, Michael Jackson was booked and released on bail after voluntarily returning from Las Vegas to California to face charges of 'lewd or lascivious acts' with a child under 14. The police mug-shot was immediately released world-wide and created a sensation as Jackson's face had not been seen unmasked in public for a number of years.

p. 14
André Kertész, **The Puppy**, 1928
André Kertész (1894–1985) enjoyed a career as a photojournalist which spanned 70 years. Born in Budapest in 1894, he moved to Paris in 1925 and became an American citizen in 1944. He loved to photograph scenes of everyday life, using small-format cameras to capture unexpected views. He never lost his childlike curiosity and freshness of vision. His achievement was to identify and freeze-frame the telling moment and the overlooked but expressive details of his subjects.

The critic and writer Roland Barthes uses *The Puppy* at the end of his meditation on photography, *Camera Lucida*, where he describes the difficulty of capturing a person in a photograph. Like Kertész, Barthes felt this could only be achieved in the revelation of the small, oblique but expressive details of the ordinary moments in life.

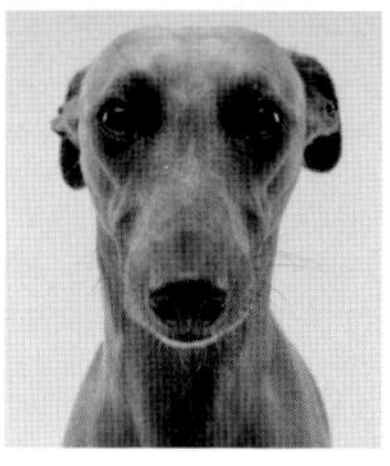

p. 15
Jo Longhurst, **Terence**, 2003
Longhurst's work explores the relationships between animal and human portraits.

When looking at a photo of a dog, we may identify with it instantly, without necessarily knowing why. Dogs are effective visual communicators precisely because they can't talk. We're used to looking closely at them to see what it is they want to tell us. A dog portrait is somehow more pared down than a similar human portrait. A human portrait, however stripped down in style and emotion, still gives us an idea of the sitter's identity. If we look at Longhurst's portrait of *Terence*, he sits looking directly into the lens with a simple grey/green backdrop reminiscent of institutional photos. The artist has made this image as if it were a human portrait, hoping to create an intimate study that shows both the character and vulnerability of the dog. Her intention was to photograph him in a format we humans are familiar with. Terence has no social mask and returns our gaze. His look is steady, yet it is immediately clear to us that his thoughts are beyond our reach. Despite the invitation in his eyes, it is impossible to know what he is thinking or feeling.
(Adapted from Jo Longhurst, 'A Brief Introduction to Dog Theory', 2003, by permission.)

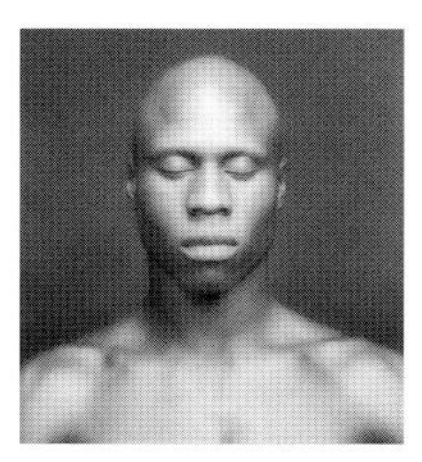

p. 20
Robert Mapplethorpe,
Ken Moody, 1983
The photographer Robert Mapplethorpe is known for his statuesque nudes, self-portraits and sensitive formal portraits of artists, actors and musicians.

Ken Moody, with his exemplary physique, was one of Mapplethorpe's favourite models. This is not a portrait of a man but a homo-erotic study of male beauty. We do not learn about Moody's personality but rather examine and admire his physical perfection. The viewer is detached, the model's face is at rest, expressionless, almost a death mask. Mapplethorpe chooses models who conform to classical canons of physical beauty, and he uses carefully controlled studio lighting to emphasise their heroic musculature and its latent power. Mapplethorpe's use of black and white lends his work a timelessness and suggests a debt to the traditions of ancient Greek and Renaissance marble statues.

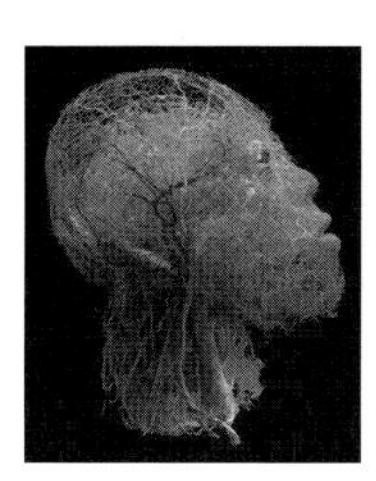

p. 22
Gunter von Hagens,
Plastinated vascular system of the head, 2000
Dr von Hagens invented the plastination process that preserves elements of the human body with reactive polymers enabling the extraction and display of the intricate workings of the human anatomy. The brain requires up to one fifth of the body's oxygenated blood and the arteries visible here supply the vital flow to the brain and the face. Von Hagens' *Body Worlds* 2004, which exhibited bodies preserved by plastination as works of art, divided public opinion.

p. 26
Edwin Romanzo Elmer,
Mourning Picture, 1890
In America in the mid-nineteenth century, as part of the tradition for posthumous mortuary portraits, it became fashionable to portray the dead as if they were alive. In previous centuries dead children were painted lying in bed. In this painting, the artist and his wife, in mourning dress, are shown seated in front of their house. Effie, their deceased daughter, stands to the left of the canvas, surrounded by her pets and toys. Posthumous commemorative paintings like this one were usually commissioned by family members as private mourning objects, and hung in semi-public spaces such as parlours.

p. 27
Unknown, **Photograph of father with mother holding dead child**, circa 1850s–1860s
In the tradition of posthumous mortuary portraits, it was more difficult for photographers than for painters to present an acceptable likeness of a dead person. Even the illusion of slumber was sometimes hard to portray. In 1855 Nathan Burgess wrote a comprehensive account of photographing the dead, 'Taking Portraits after Death':

If the portrait of an infant is to be taken, it may be placed in the mother's lap, and taken in the usual manner by a side-light, representing sleep . . .

By making three or four trials, a skilful artist can procure a faithful likeness of the deceased, which becomes valuable to the friends of the same if no other has been procured when in life.

This photograph records the desire to have a last family image. The baby is cradled in the mother's arms as if asleep.

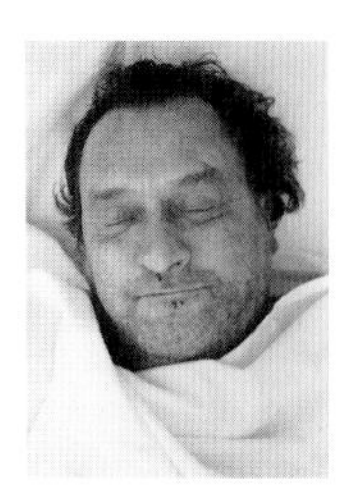

p. 28
Rudolf Schaefer,
from the series **Dead Faces**, 1986
By contrast with the unknown nineteenth-century family portrait on page 27, what is extraordinary about this portrait is how natural it appears in the context of contemporary photography that is often highly composed and manipulated.

Schaefer explores the possibility of capturing a true representation of a living person when they are literally changing second by second. When we die, we are finally still. Family and friends are usually asked whether they would like to view the deceased or if they would prefer to remember them as they were in life. There is often a fear that the dead body might be unrecognisable, which Schaefer believes is not the case. Schaefer wants to show the dignity that comes with death, to break the taboo which surrounds it.

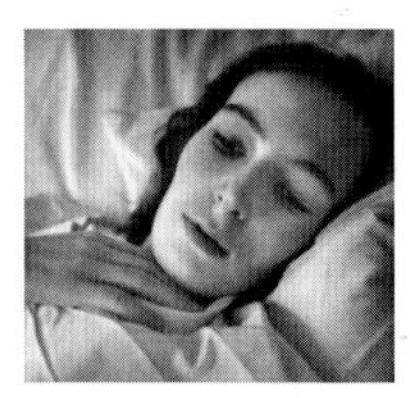

p. 29
Prof. Dr H. Killlian, **Sarkom des Schienbeines mit Metastosen in der Lunge**, 1934
This remarkable photograph is taken from a series published by Dr Hans Killian in his medical textbook *Facies Dolorosa. Das schmerzensreiche Antlitz. Physognomik und Mimik kranker Menschen*. The book is a study of pain as it is visible on the faces of the sick and is illustrated with sensitive portraits of his patients suffering from various diseases. These extraordinary images form a moving record of disease and the doctor/patient relationship. Such intimate pictures could not have been taken without mutual trust. The patients are dignified in their suffering and sometimes disturbingly beautiful. Taken at the bedside with a simple Rolleiflex camera in natural light, these black and white snapshots are intended as teaching aids. Killian believed that through close observation the doctor could learn to read the pain on the sufferer's face in order to diagnose the cause of their suffering, physical or mental, and to prescribe a cure. He felt this was an ancient but neglected art.

p. 32
Brassempuoy Venus, circa 30,000 BC
This miniature head, carved from ivory, measuring 3.5 centimetres and found at Brassempuoy, Landes, France, is one of the most famous examples of prehistoric art. It may be 30,000 years old. It is one of the few known Ice Age figures with facial features and a detailed hairstyle. The face's lionesque appearance and its crocodile-scaled hair suggests it is a symbolic image of a female figure of power. The face has no mouth and no pupils to the eyes. The depiction of a hood over the brushed-back hair has led some archaeologists to name this statue 'the hooded woman'.

p. 32
Fruitstone amulet, late 19th century
Amulets are associated with charms and talismans and bring luck to their owners. This amulet is in the form of a face and belonged to the Bushongo tribe, also known as Bakuba kingdom, in Zaire, who are noted for their crafts and fabrics. It is carved from fruitstone and measures just 4 centimetres long.

p. 36
Camille Silvy,
Adelina Patti as Martha, 1861
Introduced by French photographer A. A. Disderi in 1854, the carte de visite became a craze in the 1860s, led by Queen Victoria. These small cards with images of friends, family and celebrities were collected and treasured, and were usually kept in albums in the parlour. This is a characteristic example of a carte de visite. Subjects posed formally in their best clothes against standard studio backgrounds of balustrades, curtains and columns.

From his studio in Porchester Terrace, photographer Camille Silvy (1835–1910) produced cartes de visite for the royal family, the upper classes and the famous. The subject here is Adelina Patti (1843–1919), a celebrated soprano, who became an international opera star. Silvy produced several cartes de visite for her, some within the roles of the operatic heroines she portrayed.

p. 37
Patricia Piccinini,
Psychotourism, 1994–5

For more than a decade Australian artist Patricia Piccinini's work has examined the hopes and fears with which we greet reproductive technologies. Piccinini's work explores the plasticity of the human face and body, particularly its potential for manipulation and enhancement through bio-technological intervention. 'Give your children a chance in life: don't just leave it up to nature,' reads a caption in Piccinini's 'biosphere'.

In *The Mutant Genome Project*, Piccinini created what she described as the world's first commercially designed baby, LUMP (Lifeform with Unevolved Mutant Properties):

> LUMP is the human form completely re-engineered by an advertising agency . . . physiognomically efficient and marketably cute . . . LUMP is not only about IVF, genetic research and medical marketing, it is also about an engagement with popular and media culture and an aesthetic that admits its fascination with the plastic world of the late twentieth century. My work comes from a position that acknowledges a desire for the shiny stuff that consumer culture has to offer (plastic, TV, sneakers, the FACE).

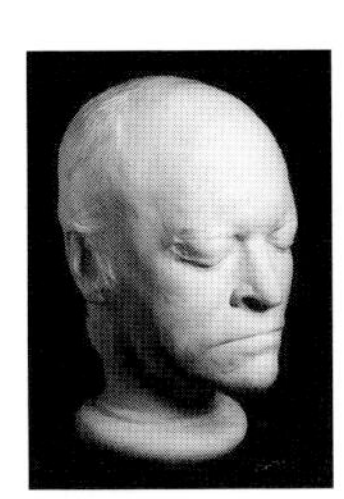

p. 46
James Deville,
William Blake, 1823

When he had this lifemask taken in his old age, poet and artist William Blake (1757–1827) was a cult figure among a small circle of Romantics. He was to become infinitely more famous after his death. In September 1823, at the age of 53, he allowed the sculptor James Deville (1776–1846) to immerse his head in plaster, with only a straw to breath through as it solidified. Before photography, masks from moulds of living and recently dead faces were the most accurate way of preserving someone's likeness. Deville probably learned the technique from his master, the sculptor Joseph Nollekens. Deville was a prominent practitioner of phrenology – reading character from the size and shape of the skull, as devised by J. Spurzheim. Deville built up a huge collection of casts and wished to include Blake's 'as representative of the imaginative faculty'.

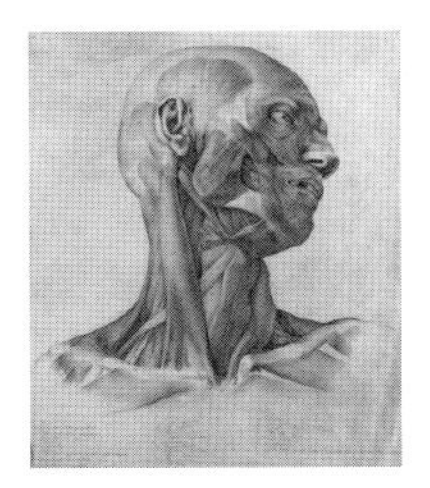

p. 48
Antonio Durelli, **Muscles and tendons of the head and neck**: écorché figure, 1837

This drawing of a head is one of a series of drawings in red chalk and pencil Durelli made of different parts of an écorché figure, a cadaver from which the skin has been peeled back to show the muscles and tendons beneath. Ecorché figures, skeletons and partially dissected bodies were traditionally depicted as if almost alive. Ecorché figure models were used as teaching aids in art academies in the nineteenth century. The study of anatomy was an essential part of an art student's education right up until the last century.

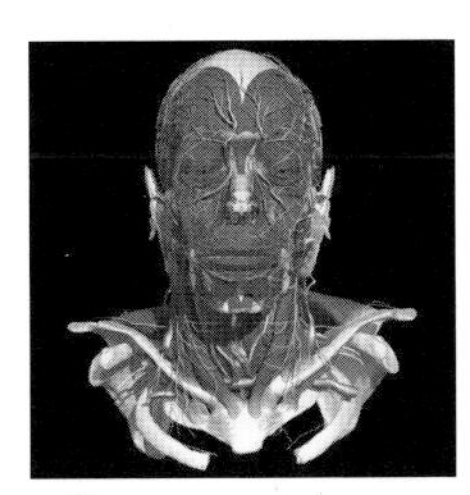

p. 49
Primal Pictures, **Interactive head and neck**, 2003

Primal Pictures worked for over ten years to create a medically accurate 3D model of the human anatomy, which it completed in June 2003. The project has also included the development of computer-generated animations showing the intricate workings of the human body in motion, to help medical practitioners and students with their training, research and diagnostic work. This example shows a fully-fleshed head with details of the intricate muscles, blood vessels and bones that create the face and its expressions.

p. 52
Francis Bacon, **Study for Portrait (after the Life Mask of William Blake)**, 1955
Francis Bacon (1909–92) kept his own copy of William Blake's lifemask in his bedroom, next to treasured personal photographs. For Bacon, Blake is the epitome of the Romantic visionary artist with the power to transcend the material world. Bacon made a series of paintings of Blake's lifemask in the mid-1950s.

The title *Study for Portrait* suggests that this is a work in progress, an attempt to get at the essence of what a portrait is. Bacon struggles with the possibility of representation and questions the nature of portraiture. How can an exact cast-copy of the features or a photograph be a truer portrait of a personality than a painting or a poem? Bacon takes inspiration from the plaster cast and transforms it into tortured human flesh. With his animated portrait, Bacon throws into relief the shortcomings of both life-casting and photography which kill what they capture. Bacon distills the very essence of this mystic visionary artist. This is a portrait not of the body, but of the soul, an electrifying vision of a disembodied spirit.

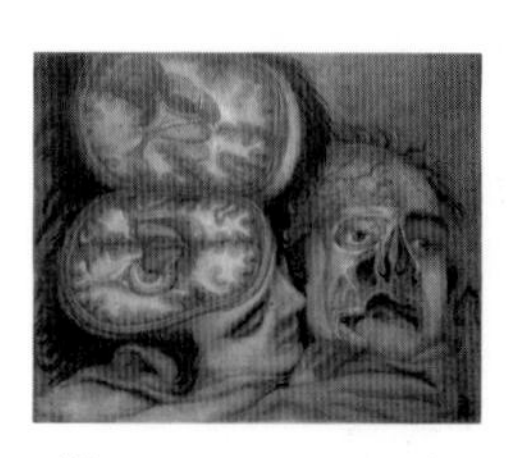

p. 53
(Jacques) J. F. Gautier d'Agoty, **Two heads with brains exposed** from *Anatomie de la tête en tableaux imprimés*, 1748
J. F. Gautier (1717–55) developed the revolutionary process of colour mezzotint which had been invented by his master, Jacob Christoph Le Blon. He initially worked with two trained anatomists, but, like many of his artist contemporaries, eventually performed his own dissections. In his book, *Myologie*, he described himself as 'demonstrator, artist and engraver all in one'. His engravings are both works of art in themselves and functional anatomical diagrams for instructing medical students.

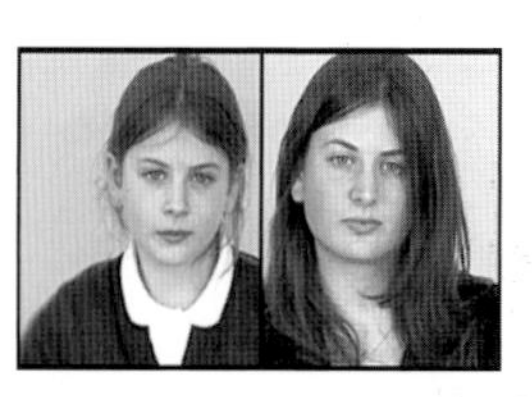

p. 54
Christian Dorley-Brown, **15 Seconds**, 1994–2004
This video was specially commissioned for the 'Future Face' exhibition. The piece consists of two 15-second informal video portraits of each participant, shown side by side, the first taken in April 1994 at age 10, the second exactly a decade later in April 2004. Dorley-Brown asked the participants to express themselves entirely through facial expression.

When the artist returned to try and find and photograph the original participants again, only eight of the young people remembered the first portrait being made, but many have now promised to return in 2014 and repeat the process. This on-going project is an eye-opening social documentary and an elegiac meditation on the passage of time and the loss of childhood innocence.

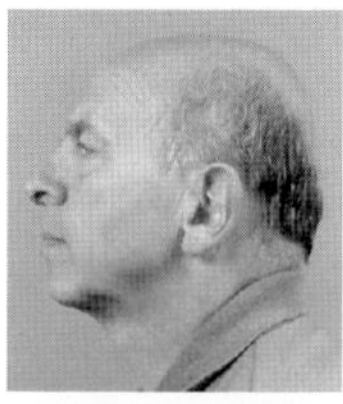

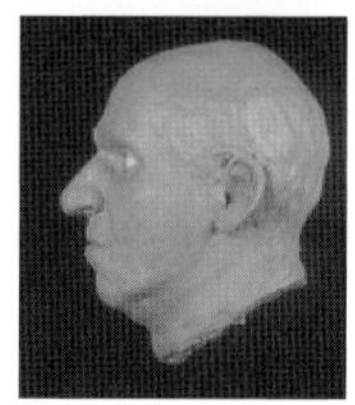

p. 55
Richard Neave,
Professor Peter Egyedi, 1996
Richard Neave, **The reconstructed head of the unknown Dutchman**, 1996
In 1996, Dr Wittkampf at Utrecht wrote to Richard Neave: ' One of the members of the scientific committee of our association has a typically shaped skull and is willing to go into the CT-scanner, so we are able to make a three-dimensional model of the skull of this colleague... Maybe it's a challenge for you to recreate [his] face without having met him.'

The result Neave achieved was remarkably close.

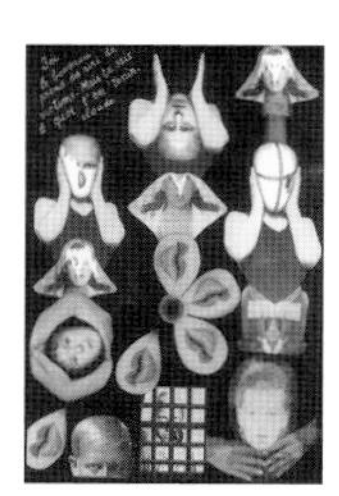

p. 64
Claude Cahun, **MRM (Sex)**, circa 1929–30
French writer and photographer Claude Cahun is known today for her startlingly contemporary photographic self-portraits and photomontages.

A contemporary of the Surrealists, she eschewed the ideals of femininity of her day by shaving her hair and dressing androgynously. She stares confrontationally out of her enigmatic self-portraits, assuming myriad disguises. Sometimes she appears as a two-headed monster, her own double or as an empty mannequin.

Through assuming and discarding the masquerade of femininity, Cahun shows us how what we think of as femininity is a culturally conditioned construct. She goes further and suggests that identity itself is a series of masks we assume. For Cahun there is no essence beneath the surface. The core is an absence and identity is a series of roles, not an inherent truth. Rediscovered in the 1990s, Cahun's work seems to anticipate artists of the 1970s such as Cindy Sherman.

p. 65
Elia Alba,
Doll Heads (Multiplicities), 2001
Elia Alba's work makes us think about our relationship with our own and other people's faces in our daily lives rather than in the isolated context of a photograph or portrait. Alba transfers photographs of real faces on to muslin, sewing and stuffing them like pillows to make dolls' heads. They are disembodied, yet lifelike in scale and appearance, poignant and unnerving by turns. In one video sequence they are seen in the sea with waves washing over them and then picked up and wrung out by a passer-by.

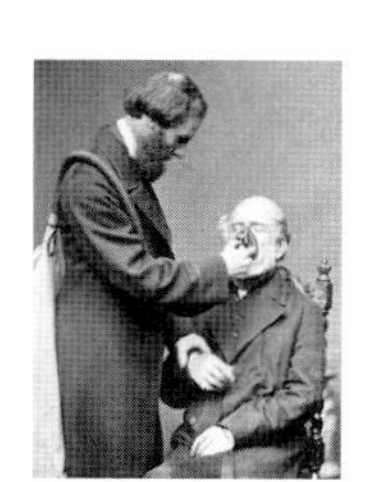

p. 74
Chloroform mask, 1862
This photograph displays J. T. Clover demonstrating his apparatus. It administered chloroform from a reservoir and through a network of tubes. He was working in the 1860s and this photograph was taken in 1862.

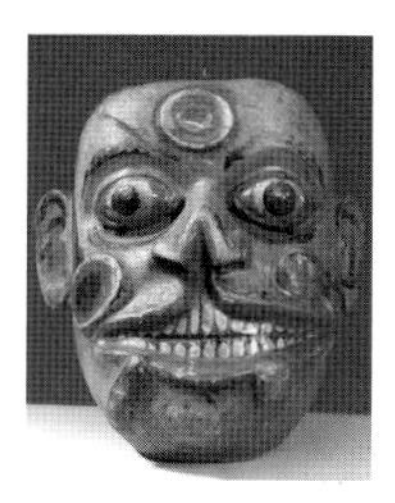

p. 82
Leprosy demon mask,
late 18th–19th century
This Sinhalese wooden mask represents Heraya, the soldier from the kolam play, a popular form of folk theatre. These dramas used masks extensively, which were based on real characters but with exaggerated features. This mask was worn for healing rituals and to get rid of evil spirits which were believed to cause illness. It shows the character's face covered with sores and leeches , indicating leprosy.

p. 83
Iron mask
This fierce iron mask with moulded eye-brows, a grotesque nose piece and furrowed lines in its forehead would have been worn by an executioner. It is European and would have been made or used between the sixteenth and seventeenth century.

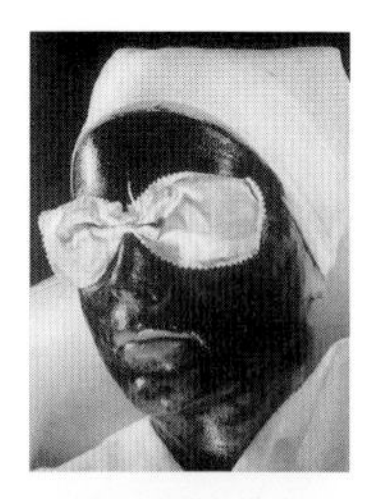

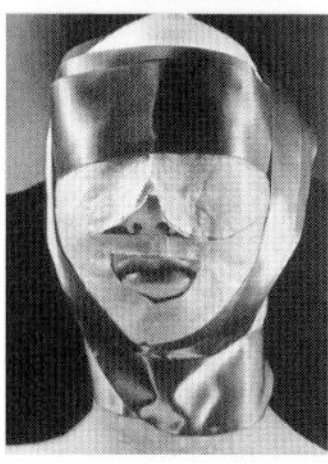

p. 88
Rubinstein Pomade Noir masque,
Condé Nast, March 1939

p. 89
Rubinstein masque, Condé Nast, 1939
Helena Rubinstein, along with Elizabeth Arden, helped to develop the cosmetics industry as we know it today. Rubinstein pioneered the creation of beauty treatments such as the face pack; these masques were intended to replenish and nourish the facial skin, restoring its youthfulness and thus equating beauty with health.

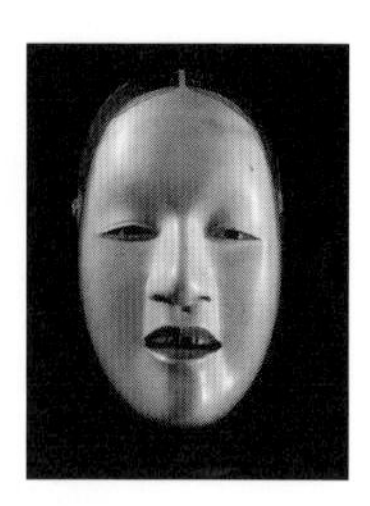

p. 94
Nōh mask of a young woman,
18th–19th century
The mask has always been a potent force in theatrical performance and carries with it a double notion of hiding and transforming identity. The mask obscures the flexibility of the face, its uniqueness and its full expressive range.

The masks of Japanese Nōh drama are generally neutral in expression, and it is the skill of the actor which brings the mask to life. The female mask, with its unearthly reflection of profoundly internalised joy and sorrow, is at the heart of the Nōh and is treated as containing a spirit of its own. A skilfully carved mask like this one will appear to have subtle changes of expression depending on the way in which the wearer turns his head. This is one of several variations of a young woman mask based on an original design by mask-maker Zeami.

p. 95
Leiter of Vienna, **Bakelite phantom for practising eye operations,**
circa 1860–90
Surgeons practised complex operations on models fitted with technology designed to emulate the human body. Materials such as Bakelite were used; in this case the resistance of the eyes was mimicked by the use of springs.

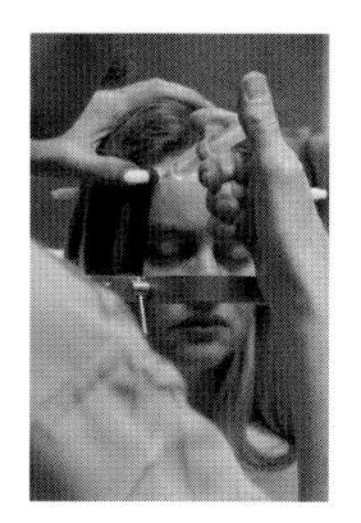

p. 96
Corinne Day, **Me just before brain surgery, London Hospital 1996,** 2000
Corinne Day entered the fashion world as a model but started taking fashion photographs for magazines in the early 1990s. She rejected the reigning fakery of airbrushed glamour and launched the waif look with her 1993 photographs of Kate Moss. These images of a bare-faced, skinny 14-year-old Moss captured the zeitgeist and transformed the face of fashion, earning Day notoriety along the way.

In her quest for authenticity, Corinne Day turned her camera on herself and her friends. A Nan Goldin for her generation, her *Diary* (2000) is a moving and uncensored journal of the lives and loves of her companions. Cruel and tender, her snapshots often depict drug-taking and sexual encounters but do not feel voyeuristic because of her emotional involvement with the subjects.

In 1996, at the age of thirty, Day was diagnosed with a brain tumour. Determined to record the intensity of the experience, she asked boyfriend Mark Szaszy to document it with her camera. This is one of the series of photographs which form an unflinching account of her treatment for the life-threatening disease.

p. 97
Max Factor Beauty Calibrator, 1932
In 1932 Max Factor unveiled his latest creation, a futuristic-looking device called 'The Beauty Calibrator', which was reputed to have the capability of measuring good looks to the hundredth part of one inch. The Beauty Calibrator was fitted with tiny thumbscrews that adjusted flexible metal bands. These bands pressed gently and closely to the contours of the wearer's face, head and neck. The object of the calibrator was to reveal how a person's facial measurements differed from those of 'the perfect face'. That determined, corrections could be made with cosmetics to give a 'perfect' illusion.

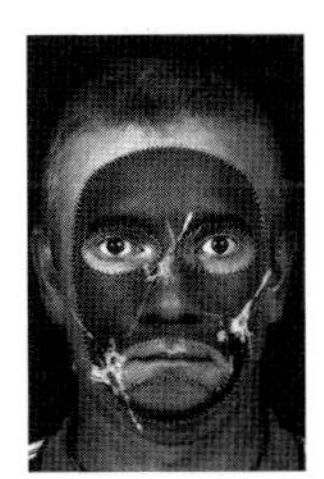

p. 100
White Man Black Mask
This image, which appears in *National Heritage, 2000* and *Colour Separation, 1996*, is a multi-platform artwork that merges more than 100 ethnically diverse photographic close-ups into eight images of four distinct racial types in both male and female variants. The eight stereotyped racial images wear masks of the other stereotypes and these masks bear the traces of racial abuse in the form of spit. The aim is to get under the viewer's skin and confront them with the power of racial stereotypes.

p. 105
Allegory of Vanity, 18th century
Vanitas is the Latin word for vanity and symbolises both the fleeting nature of life and the need to live the present moment fully. This model depicts a woman's head simultaneously fully fleshed and after death – one side of the skull is rotting. The Latin verse at the bases of the model translates as 'Emptiness, emptiness, says the speaker, emptiness, all is empty.'

p. 108
The Physiognomist, G. E. Madeley after G. Spratt, 1831
This delightful composite print depicts a physiognomist whose body is entirely made up of human faces, sitting at a table diagnosing people's physiognomic characteristics with the help of a book. Its scientific and public popularity is evidenced in this quirky print.

p. 113
Known militant suffragettes: police identification cards, 1914
The first 'portraits' of the suffragettes were variants on police mug-shots. In July 1914, Annie Hunt – a women described in contemporary newspaper accounts as 'of refined appearance and very respectably dressed' – took a butcher's cleaver to the National Portrait Gallery's portrait of Thomas Carlyle by Millais. As a result, and in an effort to protect the gallery's collection from further attacks, identification portraits of 'known militant suffragettes' together with their typed descriptions were issued by the police and distributed to the National Portrait Gallery's attendants.

p. 120
Tibor Kalman, **Black Queen Elizabeth**, Benetton *Colors* Magazine (no. 4: Race), 1993

Under the directorship of political activist Tibor Kalman in the 1990s the United Colors of Benetton magazine wore its social conscience on its sleeve. *Colors* showcased provocative images.

An issue of the magazine devoted to racism featured a fantasy series of full-page manipulated photographs imagining famous people as different races: Queen Elizabeth and Arnold Schwarzenegger as black; Pope John Paul II as Asian; Spike Lee and Michael Jackson as white. This familiar image of Queen Elizabeth II was made into an icon by Andy Warhol and it is initially shocking to see the Queen racially transformed. The artist boldly invites us to imagine a colourful future where racial stereotypes are a thing of the past.

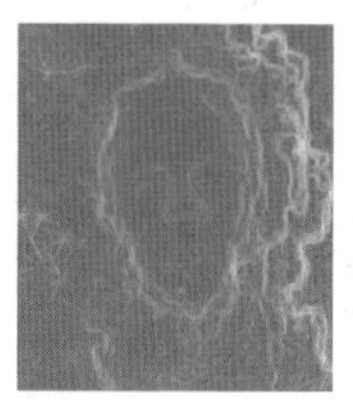

p. 129
Heather Barnett,
One Man's Land, 2002

One Man's Land is a contour map of a young man's face etched into steel. Generated from stereo images using 3D mapping technology, the information held within the portrait is accurate and unique to the sitter.

This work was produced in 2002 during contemporary visual artist Heather Barnett's residency at the satellite and aerial imaging company Infoterra to create works which explore the fascinating and technically demanding world of remote sensing and earth observation. Working in collaboration with Infoterra staff, Barnett developed projects which examine things in close-up, exploring the relationship between humanity and the land we inhabit, whether it be symbiotic, political or precarious. All her works reflect our relationship with land in some way and were produced using Infoterra's mapping and aerial imaging technologies – including thermal film, stereo photography and height elevation software.

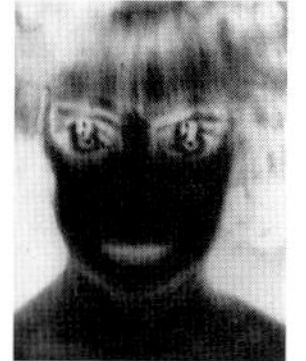

pp. 142–3
Man Ray, **Marquise Casati**, 1922

The photographer Man Ray was associated with the Dada movement and later the Surrealists in Paris. He was influenced by the erotic writings of the Marquis de Sade and used his innovative photographic techniques to refine a Surrealist iconography of the female form in his dream-like images. This is his hypnotic portrait of the beautiful and eccentric society heiress, Marquise Luisa Casati. For this photograph Man Ray had rejected the negatives because they were blurred but Casati insisted on seeing them. Man Ray recounts:

I printed up a couple on which there was the semblance of a face – one with three pairs of eyes. It might have passed for a Surrealist version of the Medusa. She was enchanted with this one – said I had portrayed her soul.

p. 146
Wanda Wulz, **Cat and I**, 1932

Italian photographer Wanda Wulz (1903–84) came from a family of photographers in Trieste. She trained with her father and took over his studio when he died, when she was 25. She started producing avant-garde portraits and technically demanding photomontages. *Cat and I* is an uncanny melding of the artist's face with a cat's and was produced by superimposing negatives. This fascinating image seems to offer us photographic evidence that the magical creatures we are familiar with from fairy tales and myths really exist. Wulz had joined the Futurists in 1931 after meeting their charismatic leader Marinetti and exhibited at Futurist photography exhibitions in the 1930s.

pp. 148-9
Douglas Gordon, **Monster I**, 1996–7
Monster is a double self-portrait. On the left half of the photograph the artist stands impassively in a white shirt, on the other his figure is repeated but his face is grotesquely disfigured. The skin has been stretched and distorted with sellotape, making Gordon's face ugly and frightening but also comic. Gordon seems to be staging an uncanny encounter with his own dark side, his evil twin, his doppelganger. Culturally, the ugly has been associated with evil: is this Gordon's own portrait of Dorian Gray? Gordon is fascinated with dualities; the tense coexistence of good and evil, life and death, light and dark.

p. 153
Giambattista della Porta,
Biggest head, 1586
Giambattista della Porta, a sixteenth-century natural philosopher, travelled Europe extensively and developed an interest in the underlying rational order in the natural world. In his book *Natural Magick* (1558) he attempted to deduce the nature of this order through theoretical and physical experimentation. Della Porta illustrated his theories with diagrams in which he used the faces of animals as a point of reference to compare the scale of their features and thus their position in this natural order.

p. 156
Jocelyn Wildenstein
This Swiss-born socialite married Alec Wildenstein, heir to a $10 billion art fortune. Alec had a 66,000 acre estate in Kenya and a love of wild animals. When their marriage faltered, Jocelyn used plastic surgery to try to transform herself into one of the big cats her husband loved. She had her lips enlarged, her face pulled back at the eyes to lend it a feline look and her skin pigment darkened. Since this extreme cosmetic procedure she has been variously dubbed 'Tiger Woman', ' The Queen of the Jungle' and 'the Bride of Wildenstein'.

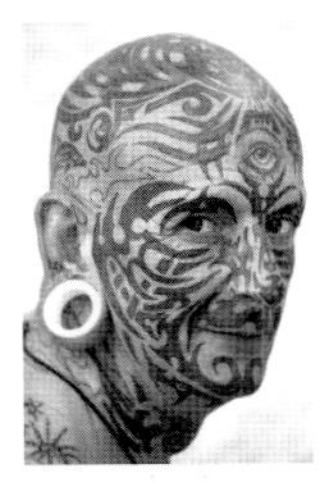

p. 158
Tattooed face, Paul Morris, 1999
The skin is a slender barrier separating the self from the outside world and has always been used as a surface for artistic expression. Since pre-history, people around the world have decorated their skins in endless ways, from body-piercing, scarification and body-painting to tattooing. Self-decoration has many functions; for celebrations, festivals, rituals and rites of passage, to show the inter-relation of the individual and society, to demonstrate creativity. Tattooing is an ancient art and its purpose varies according to the culture and place. A tattoo can indicate membership of a group, social status, criminality and more recently rebellion. A tattoo always tells a story. The word tattoo comes from the Tahitian root 'tatau' which means to inflict wounds. Tattoos are created using a hot needle to insert ink under the skin, leaving a permanent mark. Captain Cook was responsible for popularising tattoos in Britain on his return from Polynesia in the late eighteenth century. In Europe and America today colourfully dyed hair, tattoos and body-piercing are socially acceptable and have become mainstream. Twenty years ago such embellishments were considered shocking and provocative.

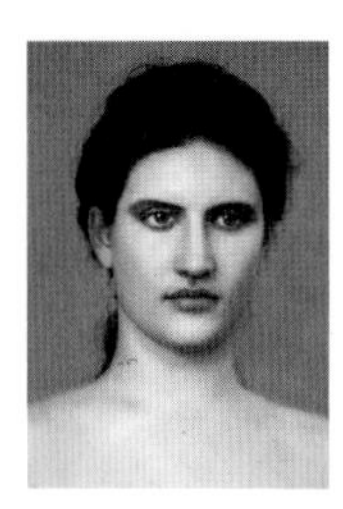

p. 159
LawickMüller (Fredericke van Lawick/Hans Müller), **Athena Velletri – Nina**, from the series **PERFECTLY SuperNATURAL**, 1999
In this series of images the perfectly symmetrical features of Greek gods and goddesses are superimposed on to human models. The work of LawickMüller shakes our deepest-held belief that the face encodes the truth about character. This image is disturbing because it does not seem so very far-fetched. These artists explore the ways we are manipulated by the media and destabilise the idea that a photographic representation of a person can tell us anything about them. This image makes explicit the process that is applied to every glossy photograph of a celebrity in fashion magazines. The artists also invite us to consider whether some bland generalised ideal of beauty is really more attractive than a face with a spark of personality.

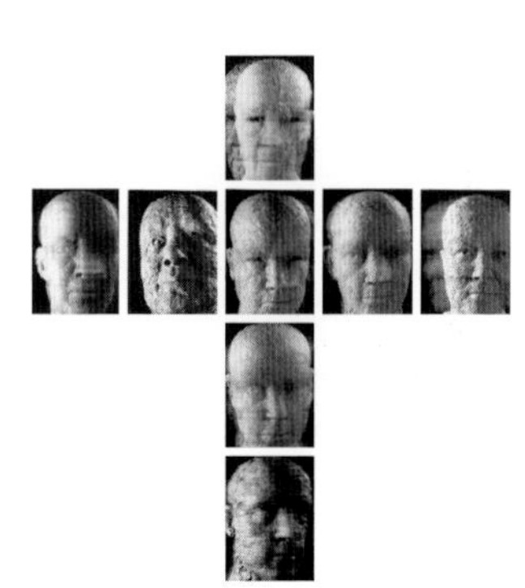

p. 164
David Hopkinson and Shelley Wilson, **Beneath the Mask**
This Sciart collaboration explores relationships amongst the invisible genetic codes and visible facial features in Shelley Wilson's family, which is of Anglo-Chinese heredity.

Hopkinson and Wilson have employed a scanning instrument that passes a narrow beam of light over a face and records its path in 3D on to a computer screen. Tiny variations in the details of a face are revealed and the information colour-coded. Thus captured, the face can be closely analysed. The family tree shows Wilson's mother, four siblings, two children and herself.

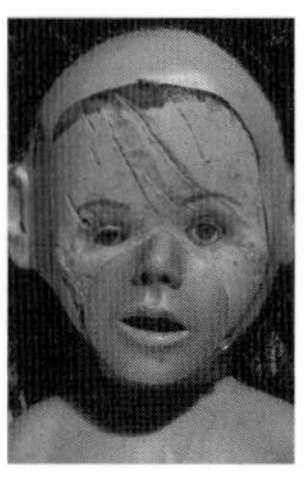

p. 165
Cindy Sherman, **Untitled MP 316**, 1995
Throughout her career Cindy Sherman has used herself as a model. In her celebrated *Untitled Film Stills* series in the 1970s, Sherman disguised herself as various heroines from fictional 1950s B-movies. She critiques the way we read a woman by her appearance. Through repeatedly reconstructing her identity and then showing these constructions to be false (we actually see the shutter release cord in her hand) she demonstrates that the masquerade of femininity can be assumed and worn like a mask. After exhausting cinematic clichés, Sherman raided historical portraits and dressed herself as Renaissance nobility, but with the addition of startling prosthetic breasts and false noses. She has also subverted fashion-shoot and centrefold genres with menacing self-portraits. In the 1990s Sherman's work became increasingly disturbing as prosthetic limbs and plastic genitals began to replace her body.

p. 196
Elastick Skin Man, 19th century
Hyperelastic skin is most often seen in people with Ehlers-Danlos syndrome, a group of rare genetic disorders that affect the production of the fibrous protein collagen. People with the syndrome are sometimes referred to as rubber men (or women) because of the increased elasticity of their skin and hyperextensible joints (the joints can be bent more than is normally possible). Elastick Skin Man's face was particularly interesting as it distorted beyond recognition.

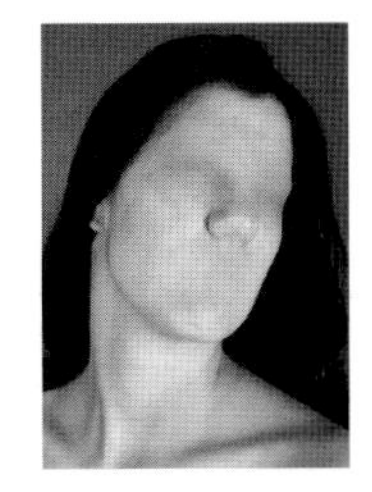

p. 197
Anthony Aziz and Sammy Cucher,
Dystopia, Maria (1994)
Pioneers in the field of digital imaging, Aziz and Cucher present us with a nightmarish vision of the future. Their *Dystopia* images represent human beings whose faces seem to have mutated, causing the skin to have grown over the features. Portraits such as *Maria* are created by erasing the subject's eyes, ears, nostrils and lips and substituting visual grafts using Photoshop.

These aberrant images are horrifying. We are trained from the moment we are born to read faces, but these blanks deny any attempt to project on to them. The lack of the eyes, the so-called windows of the soul, with which we can express so much, is especially uncanny. These pitiful image-creatures are trapped within themselves.

The artists warn of the dangers of over-reliance on information technology and the internet, and of continuing to embrace uncritically ever-accelerating technological advances.

text notes

introduction

1 Milan Kundera, *Immortality*, trans. Peter Kussi (London: Faber and Faber, 1991), pp. 35–6.

2 Charles Bell, *Essays on the Anatomy and Philosophy of Expression* (London: John Murray, 2nd edn., 1824), p. xvii.

3 *Philippe Halsman: A Retrospective* (Bullfinch Press. Little, Brown, 1998).

face to face

chapter 1
image

1 Rainer Maria Rilke, *The Notebooks of Malte Laurids Brigge*, introduction by Stephen Spender, translation John Linton (Oxford: Oxford University Press, 1984), p. 6.

2 See Select Bibliography for books on the face.

3 Marcia Pointon, 'Kahnweiler's Picasso; Picasso's Kahnweiler', in Joanna Woodall (ed.), *Portraiture: Facing the Subject* (Manchester and New York: Manchester University Press, 1977), p. 190.

4 Allan Sekula, 'The Body and the Archive', *October*, 39 (Winter 1986), pp. 3–64.

5 George Orwell, *Nineteen Eighty-four*, 1949 (Harmondsworth: Penguin, 1983), p. 230.

6 Quoted from the first edition, translated by Constance Garnett, in *The Overcoat and Other Stories* (New York: Knopf, 1923), p. 243. For the portrait in literature, see Theodore Ziolkowski, *Disenchanted Images: A Literary Iconography* (Princeton, NJ: Princeton University Press, 1977), ch. 3; A. S. Byatt, *Portraits in Fiction*, (London: Chatto and Windus, 2001).

7 For the history of facial reconstruction, see John Prag and Richard Neave, *Making Faces: Using Forensic and Archaeological Evidence* (London: British Museum Press, 1997).

8 See Susan Walker and Morris Bierbrier, with Paul Roberts and John Taylor, *Ancient Faces: Mummy Portraits from Roman Egypt* (London: British Museum Press, 1997).

9 See 'Putting Faces to the Names: Social and Celebrity Portrait Photography', in Peter Hamilton and Roger Hargreaves, *The Beautiful and the Damned: The Creation of Identity in Nineteenth-Century Portrait Photography* (London: The National Portrait Gallery in association with Lund Humphries, 2001), pp. 17–54.

10 See Richard Brilliant, *Portraiture* (London: Reaktion, 1991); Melissa E. Feldman, *Face-Off: The Portrait in Recent Art* (Philadelphia: Institute of Contemporary Art, 1996); Katherine Hoffman, *Concepts of Identity: Historical and Contemporary Images and Portraits of Self and Family* (New York: Harper Collins, 1996); Marcia Pointon, *Portraiture and Social Formation in Eighteenth Century England* (New Haven and London: Yale University Press, 1993); Robert A. Sobieszek, *The Ghost in the Shell: Photography and the Human Soul, 1850–2000. Essays on Camera and Portraiture* (Cambridge, Mass: MIT Press, 1999); and Woodall (ed.), op. cit.

11 Unpublished paper given at the National Portrait Gallery Conference on 'Faciality', November 2003.

12 Quoted in Hamilton and Hargreaves, op. cit., p. 49.

13 Roland Barthes, *Camera Lucida: Reflections on Photography* (London: Jonathan Cape, 1982), p. 81.

14 Jay Ruby, *Secure the Shadow: Death and Photography in America* (Cambridge, Mass: MIT Press, 1995), p. 179. See also William C. Darrah, *Cartes de Visite in Nineteenth Century Photography* (Gettysburg, PA: Darrah, 1981), p. 39, and Philippe Ariès, *Western Attitudes Towards Death: From the Middle Ages to the Present* (Baltimore, Johns Hopkins University Press, 1974), for historical context.

15 Quoted in Heinz K. and Bridget A. Hemisch, *The Photographic Experience 1839–1914* (Pennsylvania: Penn State Press, 1994), p. 166.

16 From *Paul Klee*, exhibition catalogue (London, Marlborough Fine Arts, 1966), no. 45, quoted in Brilliant, op. cit, p. 171.

17 Kundera, op. cit., p. 34.

18 Malvina Hoffman, *Sculpture Inside Out* (London: G. Allen and Unwin, 1939), p. 135.

19 There is a vast literature on physiognomics, but see Mary Cowling, *The Artist as Anthropologist: The Representation of Type and Character in Victorian Fiction* (Cambridge: Cambridge University Press, 1989); E. C. Evans, 'Physiognomics in the Ancient World', *Transactions of the American Philosophical Society*, 59 (1969); E. H. Gombrich, 'The Mask and the Face: The Perception of Physiognomic Likeness in Life and Art', in M. Mandelbaum (ed.), *Art, Perception and Reality* (Baltimore, MD: Sparrow, 1972); and Graeme Tytler, *Physiognomy in the European*

Novel (Princeton, NJ: Princeton University Press, 1982).

20 See Olivier Dyens, *Metal and Flesh: The Evolution of Man – Technology Takes Over* (Cambridge, Mass: MIT Press, 2001).

21 See Catherine Ikam and Louis Fléri, *Portraits. Réel/Virtuel* (Paris: Mairie de Paris, 1999), p. 16.

22 Nathaniel Hawthorne, 'The Prophetic Pictures', *Twice Told Tales and Other Short Stories* (London: Readable, 1842), p. 23.

chapter 2 bare essentials

1 Jenny Diski, *The Dream Mistress* (Guernsey, Channel Islands: Phoenix, 1996), pp. 16–17.

2 Sarah Simblet, 'The Head', in *Anatomy for the Artist* (London: Dorling Kindersley, 2001), pp. 47–57. See also Jonathan Cole, 'Bone to Brain', in *About Face*, (Cambridge, Mass: MIT Press, 1998), pp. 43–52.

3 Paolo G. Morselli, 'The Minotaur Syndrome: Plastic Surgery of the Facial Skeleton', *Aesthetic Plastic Surgery*, 17: 99–102, 1993.

4 David Sylvester, *The Brutality of Fact: Interviews with Francis Bacon*, third enlarged edition (London: Thames and Hudson, 1987), p. 50.

5 In personal correspondence, 2004.

6 Guillaume Benjamin Amand Duchenne de Boulogne, *The Mechanism of Human Facial Expression*, ed. and trans. R. Andrew Cuthbertson (Cambridge: Cambridge University Press, 1990), p. 19. See Robert A. Sobieszek, 'The Clinical Aesthetics of Duchenne de Boulogne', *The Ghost in the Shell*, op. cit., pp. 32–79, for a detailed account of Duchenne's work.

7 Paul Ekman, *Emotions Revealed: Understanding Faces and Feelings* (London: Weidenfeld and Nicolson, 2003), p. 15.

8 See Maurice Merleau-Ponty, *The Primacy of Perception* (Evanston: Northwestern University Press, 1964), p. x; Emmanuel Levinas, *Collected Philosophical Papers* (Dordrecht, Netherlands: M. Nijhoff, 1986); 'Max Picard and the face', trans. Michael B. Smith, in *Proper Names*, Werner Hamacher and E. Wellbery, eds. (Stanford: Stanford University Press, 1996), pp. 94–8.

9 Barthes, op. cit., p. 13.

10 Jonathan Cole, *About Face* (Cambridge, Mass: MIT Press, 1998), pp. 2, 192.

11 I am indebted to Andrew Bamji for introducing me to the material concerning facial surgery of World War I in the Sidcup archive and in particular to the work of Harold Gillies and the writings of Ward Muir. See Andrew Bamjii, 'In Peace and in War', unpublished lecture, 1998, which contains detailed information on the history of Queen Mary's Hospital and the development of facial surgery, and which I have summarised here.

12 Reginald Pound, *Gillies: Surgeon Extraordinary. A biography* (London: Michael Joseph, 1964), quoted in Bamjii (1998).

13 Cpl. Ward Muir, *The Happy Hospital* (London: Simpkin, Marshall, Hamilton, Kent & Co. Ltd, 1918), p. 143.

14 Carl Ferdinand von Graefe, *Rhinoplastik; oder, Die Kunst der Verlust der Nase organisch zu ersetzen, in ihren früheren Verhältnissen erforscht und durch neue Verfahrungsweisen zur höheren Vollkommenheit gefördet* (Berlin: Realschulbuchhandlung, 1818), p. vi.; English translation quoted in Sander L. Gilman, *Making the Body Beautiful: A Cultural History of Aesthetic Surgery* (Princeton, New Jersey: Princeton University Press, 1999).

15 Emma Chambers, *Henry Tonks: Art and Surgery* (London: UCL, 2002), p. 16.

16 See Gilman, op. cit.

17 Quoted in Alan Clive Roberts, *Facial Prosthesis: The Restoration of Facial Defects by Prosthetic Means* (London: Henry Kimpton, 1971), p. 1.

18 Anna Coleman Ladd, 'How Wounded Soldiers Have Faced the World Again with "Portrait Masks" ', interview by P. Kind, *St Louis Post Dispatch*, 26 March 1933; and see Claudine Mitchell, 'Facing Horror: Women's Work, Sculptural Practice and The Great War', in Valerie Mainz and Griselda Pollock, *Work in Modern Times: Visual Mediations and Social Processes* (Ashgate: 2001), for a detailed investigation of the work of Anna Coleman Ladd.

19 Cpl. Ward Muir, 'The Men with New Faces', *The Nineteenth Century*, October 1917, p. 752.

20 See Emily Mayhew's history of the work of plastic surgeon Archibald McIndoe, to be published in 2005, entitled 'The Reconstruction of Warriors'.

21 W. T. Benda, *Masks:* Illustrated by the Author. Introduction by Frank Crowninshield (New York: Watson Gupthill Publications Inc., 1944), p. 2.

22 Ibid., pp. vii–viii.

23 See Jeff Rian, 'The Artifacts of Life', on eyestorm.com for a detailed analysis of Khazem's work.

24 See Erin O'Connor, *The Love Song of Plastic Surgery*, erinoconnor.org/writing/plastics.shtml, and Ellen Feldman, americanheritage.com

25 Gaston Leroux, *The Phantom of the Opera*, introduction by Max Byrd (New York: New American Library, 1987), p. 9. See Gilman, op. cit., for extensive discussion of the figure of Erik in *Phantom*.

26 Murray Berger, quoted in Elizabeth Haiken, *Venus Envy: A History of Plastic Surgery* (Baltimore: The Johns Hopkins University Press, 1997), p. 221.

27 A more detailed account is to be found in Elizabeth Haiken and Erin O'Connor. I have summarised their arguments here.

28 Quoted in Haiken, op. cit., pp. 95–6.

29 See Gilman, op. cit., pp. 3–8.

30 See Royal College of Surgeons, 'Working Party Report on Facial Transplantation', November 2003.

chapter 3 data face

1 Edgar Allan Poe, 'The Oval Portrait' (1842), *Selected Writings* (Harmondsworth: Penguin, 1967), p. 251.

2 Stephen Jay Gould, *The Mismeasure of Man* (New York: W. W. Norton and Company, 1981), p. 174.

3 Quoted in I. Taylor, P. Walton and J. Young, *The New Criminology: For a Social Theory of Deviance* (London: Routledge & Kegan Paul, 1973), p. 325.

4 Gombrich, op. cit., p. 35.

5 Bertillon, quoted in Christian Pheline, 'L'image accusatrice', in *Les Cahiers de la Photographie No 17* (1985), p. 129.

6 Hamilton and Hargreaves, op. cit., p. 96.

7 Norman Bryson, 'Facades', in Maurice Tuchman and Virginia Rutledge, eds., 'Hidden in Plain Sight: Illusion in Art from Jasper Johns to Virtual Reality', unpublished exhibition catalogue (Los Angeles: Los Angeles Country Museum of Art, 1966), p. 51. Quoted in Sobieszek, op. cit.

8 Vicki Bruce and Andy Young, *In the Eye of the Beholder: The Science of Face Perception* (Oxford: Oxford University Press, 1998), pp. 47–84.

9 Quoted in Susan Kuchinska, 'Image Is Everything', *Wired*, 18 June 1998.

10 Claudia Johnson, 'Fair maid of Kent? The arguments for (and against) the Rice portrait of Jane Austen', *Times Literary Supplement*, 13 March 1998, p. 15.

11 Philippe Parreno and Pierre Huyghe, quoted in Arne Altena, 'Between illusion and artificiality: computer animated heads', outlineamsterdam.nl/nlprogramma/2003

12 Mark B. N. Hansen, 'Affect as Medium, or the "digital-facial-image"', *Journal of Visual Culture*, 2,2 (August 2003), p. 214. Hansen is currently working on *Humanising the Posthuman*, a theoretical study of bodily agency in the age of cognitive neuroscience.

13 Ikam and Fléri, op. cit., pp. 27–8.

chapter 4 extreme face

1 Rilke, op. cit., p. 7.

2 Johann Kaspar Lavater, *Essays on Physiognomy: Designed to Promote the Knowledge and the Love of Mankind*, trans. Thomas Holcroft, 9th ed. (London: William Tegg, 1855), pp. 10–11.

3 See Wendy Wicks Reaves, *Celebrity Caricatures in America* (Washington DC: Smithsonian, 1998).

4 W. H. Auden, 'Concerning the Unpredictable', *New Yorker*, 21 February 1970, p. 124.

5 Charles Mills Gayley, *The Classic Myths: In English Literature and in Art* (Boston: The Athenaeum Press, 1911), p. 208.

6 O'Connor, op. cit., Part Five.

7 Daniel W. Smith, Introduction to *Gilles Deleuze: Essays Critical and Clinical*, trans. Daniel W. Smith and Michael A. Greco (London and New York: Verso, 1998), p. xxxii.

8 Quoted in Alberto Manguel, *Bride of Frankenstein* (London: British Film Institute, 1997), pp. 20–21.

9 Ricardo Galeazzi, 'Rebuilding disabled soldiers: wonderful work that Italy is doing to render maimed men self-supporting', in *Current History* (New York: July 1918), p. 101.

10 Angela Carter, 'The Tiger's Bride', in *The Bloody Chamber and Other Stories* (Harmondsworth: Penguin, 1981), p. 53.

11 Marsha Bentley Hale, 'Lasting expressions', *Visual Merchandising and Store Design*, November 1985, p. 45. See also Emily and Per Ola d'Aulaire, 'Mannequins: our fantasy figures of high fashion', *Smithsonian*, April 1991.

12 Jonathan Cole, op. cit., p. 96. Donna Williams has written two books on her personal experience of autism: *Nobody Nowhere* (New York: Random House; London: Corgi Books, 1992) and *Somebody Somewhere* (New York: Random House; London: Corgi Books, 1994).

13 Al Lingis, 'The tact and tenderness of the light', unpublished paper delivered at the National Portrait Gallery 'Faciality' conference, December 2003.

14 Barthes, op. cit., pp. 63–72.

identikit face

1 A. M. Burton, V. Bruce and N. Dench (1993), 'What's the difference between men and women? Evidence from facial measurement', *Perception*, 22, 153–76.

2 V. Bruce, A. M. Burton, E. Hanna, P. Healey, O. Mason, A. Coombes, R. Fright and A. Linney (1993), 'Sex discrimination – how do we tell the difference between male and female faces?', *Perception*, 22, 131–52.

3 R. Campbell, P. J. Benson, S. B. Wallace, S. Doesbergh and M. Coleman (1999), 'More about brows: How poses that change brow position affect perceptions of gender', *Perception*, 28, 489–504.

4 P. M. Le Gal and V. Bruce (2002), 'Evaluating the independence of sex and expression in judgments of faces', *Perception & Psychophysics*, 64, 230–43.

5 I. S. Penton-Voak, D. I. Perrett, D. L. Castles, T. Kobayashi, D. M. Burt, L. K. Murray and R. Minamisawa (1999), 'Menstrual cycle alters face preference', *Nature*, 399, 741–2.

6 F. Galton, *Inquiries into Human Faculty and Its Development* (London: Macmillan, 1883).

7 A. W. Young, D. J. Hellawell and D. C. Hay (1987), 'Configural information in face perception', *Perception*, 16, 747–59.

8 V. Bruce, H. Ness, P. J. B. Hancock, C. Newman and J. Rarity (2002), 'Four heads are better than one: combining face composites yields improvements in face likeness', *Journal of Applied Psychology*, 87, 894–902.

9 P. J. B. Hancock (2000), 'Evolving faces from principal components', *Behaviour Research Methods, Instruments and Computers*, 32, 327–33.

10 A. M. Burton, S. Wilson, M. Cowan and V. Bruce (1999), 'Face recognition in poor quality video: evidence from security surveillance', *Psychological Science*, 10, 243–8.

11 V. Bruce, Z. Henderson, K. Greenwood, P. Hancock, A. M. Burton and P. Miller (1999), 'Verification of face identities from images captured on video', *Journal of Experimental Psychology: Applied*, 5, 339–360; and V. Bruce, Z. Henderson, C. Newman and A. M. Burton (2001), 'Matching identities of familiar and unfamiliar faces caught on CCTV images', *Journal of Experimental Psychology: Applied*, 7, 207–18.

12 A. W. Young, D. C. Hay and A. W. Ellis (1985), 'The faces that launched a thousand slips: everyday errors and difficulties in recognising people', *British Journal of Psychology*, 76, 495–523.

13 P. Chiroro and T. Valentine (1995), 'An investigation of the contact hypothesis of the own-race bias in face recognition', *Quarterly Journal of Experimental Psychology*, 48A, 879–94.

14 J. Daugman (2003), 'The importance of being random: statistical principles of iris recognition', *Pattern Recognition*, 36, 279–91.

conclusion

1 John Liggett, *The Human Face* (London: Constable, 1974), p. 275.

2 See Robin Dunbar, *The Human Story: A New History of Mankind's Evolution* (London: Faber, 2004).

3 Kobo Abe, *The Face of Another*, trans. E. Dale Saunders (New York: Vintage, 1966), pp. 31, 40.

select bibliography

photography

Barthes, Roland, *Camera Lucida: Reflections on Photography*, trans. Richard Howard (London: Jonathan Cape, 1982); 'The Face of Garbo', *Mythologies*, trans. Annette Lavers (London: Paladin, 1973).

Halsman Bello, Jane, and Bello, Steve, eds., *Philippe Halsman: A Retrospective*; introduction by Mary Panzer (New York and Canada: Bullfinch Press, 1998); exhibition curated by Mary Panzer at the Smithsonian National Portrait Gallery, Washington.
An exploration of Halsman's photography which seeks to bridge the gap between face and identity and looks at ways of revealing the 'inner' self.

Hamilton, Peter, and Hargreaves, Roger, *The Beautiful and the Damned: Nineteenth-Century Portrait Photography* (London: National Portrait Gallery in association with Lund Humphries, 2001); exhibition curated by Roger Hargreaves at the National Portrait Gallery, London.
The Beautiful and the Damned explores the parallel developments of celebrity and surveillance portraiture between 1860 and 1900 against the background of the nineteenth-century belief in the 'science' of physiognomy, the study of genetics and the belief-systems and aesthetics of social Darwinism.

Iles, Chrissie, and Roberts, Russell, eds., *Invisible Light: Photography and Classification in Art, Science and the Everyday* (Oxford: Museum of Modern Art, 1997); exhibition curated at the Museum of Modern Art, Oxford by Russell Roberts.
Crossing over conventional boundaries of time, space and nationality, as well as blurring the line between science and art, *Invisible Light* looks at the contradictory ways we have attempted to order the world through the photographic image from the nineteenth century to the present.

Madlow, Ben, *A Narrative History of the Face in Photography* (New York and Canada: Little Brown, 1977).
A beautifully illustrated historical account of portrait photography.

National Portrait Gallery, *Faces of the Century: A Sainsbury's Photographic Exhibition* (London: National Portrait Gallery, 1999).
A portrait of the twentieth century through 100 photographs, chosen by ten contemporary figureheads from the world of music, science, politics, business, high fashion and the arts. Illustrates the diversity of events and individuals that have transformed Britain since 1900.

Ruby, Jay, *Secure the Shadow: death and photography in America* (Cambridge, Mass.: MIT Press, 1995).
Secure the Shadow combines cultural anthropology and visual analysis to explore the photographic representations of death in the United States from 1840 to the present.

Rugoff, Ralph, with contributions from Anthony Vidler and Peter Wollen, *Scene of the Crime* (Cambridge, Mass. and London: MIT Press, 1997); exhibition curated by Ralph Rugoff at the Armand Hammer Museum of Art and Cultural Center at UCLA.
Drawing on cultural fascination with crime scenes and investigations, *Scene of the Crime* investigates aesthetic practices that address the art object as a kind of evidence, a clue to absent meanings and prior actions, and links it to a forensic approach.

Sekula, Allan, 'The Body and the archive', *October*, 39 (Winter 1986), pp. 3–64.
An examination of the advent of photography in nineteenth-century France, which pays particular attention to the ways in which the operations of the archive served as devices of regulatory control.

Sobieszek, Robert A., *Ghost in the Shell: Photography and the Human Soul, 1850–2000* (Cambridge, Mass: MIT Press, 1999); exhibition curated by Robert Sobieszek at the Los Angeles County Museum of Art.
Ghost in the Shell takes as its premise the idea that the face is a reflection of the inner self. Tracing modern photographic portraiture over the past 150 years, the book reveals the many ways the photographic arts have investigated, represented, interpreted and subverted the human face and, consequently, the human spirit.

Stafford, Barbara, *Body Criticism: Imaging the Unseen in Enlightenment Art and Medicine* (Cambridge, Mass. and London: MIT Press, 1991); and *Good Looking: Essays in the Virtue of Images* (Cambridge, Mass and London: MIT Press, 1996).
Body Criticism looks at ethics and aesthetics in the context of imaging in art and medicine. *Good Looking* contains twelve essays which meditate on the ways in which educational, medical and legal institutions are being transformed by the explosion of optical information, the urgency of inventing an imaging interdiscipline and the ethical dilemmas of technological transparency.

Svenson, Arne, *Prisoners* (New York: Blast Books, 1997).
A collection of 70 extraordinary portraits of prisoners of the end of the nineteenth and start of the twentieth centuries, and the fascinating newspaper and prisoner accounts of their day describing the crimes of which they were accused. Printed from their original glass negatives.

portraiture

Brilliant, Richard, *Portraiture* (London: Reaktion, 1991).
Portraiture investigates the genre as a particular phenomenon in Western art and in terms of its sensitivity to changes in perceptions of the individual in society. Includes an analysis of the connections between the subject matter of portraits and the

beholder's response – the response he or she makes to the image itself and to the person it represents.

Byatt, A. S., *Portraits in Fiction* (London: Chatto and Windus, 2001).
A fascinating exploration of the relationships between portraits and novels.

Pointon, Marcia, *Hanging the Head: Portraiture and Social Formation in Eighteenth Century England* (New Haven and London: Yale University Press, 1993).
England in the eighteenth century possessed a thriving portrait culture which was part of a network of visual communication that encompassed print-collecting, popular performance and figurative acts of speech. *Hanging the Head* demonstrates how portraiture provided mechanisms both for constructing and accessing a national past, and for controlling a present that appeared increasingly unruly.

Sylvester, David, *The Brutality of Fact: Interviews with Francis Bacon*, 3rd enlarged edition (London: Thames and Hudson, 1987).
Extraordinarily revealing interviews with Francis Bacon conducted by Sylvester over a period of 25 years, providing a statement of Bacon's attitude towards his own work and to painting more generally.

Time Inc., *Faces of Time: 75 Years of Time Magazine Cover Portraits*, essay by Frederick S. Voss, introduction by Jay Leno; (New York: Bullfinch Press, 1998); exhibition curated by Frederick Voss, historian and curator for the TIME Collection at the National Portrait Gallery, Washington. Celebrating *Time* magazine's 75th anniversary, this book presents 75 artworks commissioned for the magazine's covers by some of the century's best-known artists, including Roy Lichtenstein's 1968 image of Bobby Kennedy, Ben Shahn's Martin Luther King and Gerald Scarfe's caricatures of the Beatles.

Woodall, Joanna, ed., *Portraiture: Facing the Subject* (Manchester and New York: Manchester University Press, 1977).
A collection of essays which offers a history of portraiture, examining the key developments of periods progressing from the Italian Renaissance to Dutch seventeenth-century portraiture and on to Picasso, surrealism, Lucian Freud and Cindy Sherman.

masks

Gombrich, E. H., 'The mask and the face: the perception of physiognomic likeness in life and art', in M. Mandelbaum, ed., *Art, Perception and Reality* (Baltimore, MD: Sparrow, 1972).
An examination of the relations between face and mask and a critique of physiognomy.

Mack, John, ed., *Masks and the Art of Expression* (New York: Harry N. Abrams, 1994).
Illustrated with photographs of 150 remarkable objects, *Masks and the Art of Expression* is an expansive survey of the complex traditions that have determined uses and meanings of masks from archaic cultures to the present day.

Nunley, John W., and McCarty, Cara, *Masks: Faces of Culture* (New York: Harry N. Abrams, 1999).
A cultural history of the power and mystery of masks in religious ritual, warfare and entertainment from palaeolithic times to the present.

historical background

Darwin, Charles, *The Expression of the Emotions in Animal and Man* (London: John Murray, 1872).

Duchenne de Boulogne, Guillaume Benjamin Amand, *The Mechanism of Human Facial Expression*, ed. and trans. R. Andrew Cuthbertson (Cambridge: Cambridge University Press, 1990).

Kaufman, Matthew H., *Musket-Ball and Sabre Injuries from the First Half of the Nineteenth Century* (Edinburgh: Metro Press, 2003).

Lavater, Johann Kaspar, *Essays on Physiognomy: Designed to Promote the Knowledge and Love of Mankind*, trans. Thomas Holcroft, 9th edition (London: William Tegg, 1855).

Le Brun, Charles, *Conférence de M. Le Brun sur l'expression générale et particulière* (1734). Intro. Alan T. Mckenzie. The Augustan Reprint Society, nos. 200–201.

Lombroso, Cesare, *Criminal Man according to the classification of Cesare Lombroso. Briefly summarised by his daughter Gina Lombroso Ferrero. With an introduction by Cesare Lombroso* (New York and London: G. P. Putnam's Sons, 1911).

philosophy

Aristotle, 'Physiognomics', *Minor Works*, trans. W. S. Hett (London: William Heinemann Press, 1955).

Agamben, Giorgio, 'The face', in *Means Without End: Notes on Politics*, trans. Vincenzo Binetti and Cesare Casarino (Minneapolis: University of Minnesota Press), vol. 20, pp. 90–99.

Deleuze, Gilles, *Cinema 1: The Movement Image*, trans. H. Tomlinson and B. Habberjam (Minneapolis: University of Minnesota Press, 1986).

Deleuze, Gilles, and Guattari, Felix, *A Thousand Plateaus*, trans. Brian Massumi (Minneapolis: University of Minneapolis Press, 1987).

Kristeva, Julia, trans. Leon S. Roudiez, *Strangers to Ourselves* (New York and London: Harvester Wheatsheaf, 1991).

Levinas, Emmanuel, *Collected Philosophical Papers* (Dordrecht, Netherlands: M. Nijhoff, 1986); 'Max Picard and the face', trans. Michael B. Smith, in *Proper Names* (London:

The Athlone Press, 1996).

Merleau-Ponty, Maurice, *The Primacy of Perception* (Evanston: Northwestern University Press, 1964).

Schopenhauer, Arthur, Essays and Aphorisms, trans. K.J. Hollingdale (Harmondsworth: Penguin, 1970).

Wittgenstein, Ludwig, *Remarks on the Philosophy of Psychology* (Chicago: Chicago University Press, 1980).

face

Abe, Kobo, *The Face of Another*, trans. E. Dale Saunders (New York: Vintage Books, 1966).
A novel about a scientist hideously deformed in a laboratory accident – a man who has lost his face. His only means of entry back into the world is to create a mask so perfect as to be undetectable. A remorseless meditation on face and facelessness, nature, identity and society.

Bates, Brian, with Cleese, John, *The Human Face* (London: BBC, 2001).

The Human Face traces the evolutionary, social and psychological development of the human face. Divided into six chapters on origins, identity, expression, beauty, vanity and fame.

Burson, Nancy, *Faces* (New Mexico: Twin Palms Publishers, 1993).
Extraordinary black-and-white portraits of people with facial abnormalities: Burson claims that people are beautiful no matter what they look like and that distinctions between normal and abnormal don't exist.

Liggett, John, *The Human Face* (London: Constable, 1974).
An exploration of the meaning of the face and of ways of understanding and measuring its relation to self and identity.

McNeill, Daniel, *The Face* (Boston and London: Little, Brown, 1998).
An exploration of the face in evolutionary psychology, gain theory and socio-biology. *The Face* also discusses the history of mirrors and kissing, the Man in the Iron Mask, nose rings, the power of staring, clown make-up, eye spots on fish and insects, hypnotism, and the idols of Easter Island.

Young, Lailan, *The Secrets of the Face: Love, Fortune, Personality Revealed the Siang Mien Way* (London: Hodder and Stoughton, 1983).
The dictionary definition of 'Siang Mien' is 'reading faces', 'physiognomy' and 'telling destiny by inspecting your countenance'. This book outlines the Chinese art of face reading through 188 drawings and diagrams and 73 photographs. The face is examined section by section.

Walker, Susan, and Bierbrier, Morris, *Ancient Faces: Mummy Portraits from Roman Egypt* (London: British Museum Press, 1997); exhibition curated by Susan Walker and Morris Bierbrier at the British Museum.
Examination of around 200 of the painted mummy portraits of Egypt's Roman period.

skin, cosmetics and beauty

Brand, Peg Zeglin, *Beauty Matters* (Bloomington and Indianapolis: Indiana University Press, 2000).
Drawing from visual art, dance, cultural history and literary and feminist theory, the essays in this book explore the values and politics of beauty. Beauty has captured human interest since before Plato, but how, why and to whom does beauty matter in today's world?

Etcoff, Nancy, *The Survival of the Prettiest: The Science of Beauty* (New York: Doubleday, 1999).
An investigation of the 'science' of beauty and its impact on society.

Gröning, Karl, *Decorated Skin: A Survey of World Art* (London: Thames and Hudson, 1997).
Portraits with body painting, scarification, piercing and tattooing; international body art, past and present.

Lupton, Ellen, *Skin: Surface, Substance and Design* (London: Laurence King, 2002); exhibition at the Cooper Hewitt Museum, New York.
Skin not only looks at the complex membrane that holds the body together but also embraces the full spectrum of design, from product to architecture, fashion and media.

Pacteau, Francette, *The Symptom of Beauty: Essays in Art and Culture* (Cambridge, Mass: Harvard University Press), 1994.
A fascinating theoretical investigation of the place and impact of beauty in art and culture

Woodhead, Lindy, *War Paint: Helena Rubinstein and Elizabeth Arden. Their Lives, Their Times, Their Rivalry.* (London: Virago Press, 2003)
The story of feminine vanity and marketing genius. Tracks the emergent trend of beauty culture in the early 1900s, when women still painted their faces with toxic white lead and skin creams still contained arsenic.

surgery, anatomy and forensics

Bell, Charles, *Essays on the Anatomy of Expression in Painting* (London: George Bell and Sons, 1806).
Examination of the anatomy of expression and applications for art.

Chambers, Emma, *Henry Tonks: Art and Surgery* (London: UCL, 2002); exhibition curated by Emma Chambers at University College London in 2002.
An early training as a surgeon gave Henry Tonks a detailed knowledge of anatomy and the dual concerns of art and anatomy resonate throughout his work. In this critical study of his

portraits, Emma Chambers shows how Tonks sought to convey the structure of the face damaged by injury and restored by surgery.

Gilman, Sander L., *Creating Beauty to Cure the Soul* (Durham, NC: Duke University Press, 1998); *Making the Body Beautiful: A Cultural History of Aesthetic Surgery* (Princeton, New Jersey: Princeton University Press, 1999).
Examination of the cultural dimensions of modern aesthetic surgery and its history.

Petherbridge, Deanna, *The Quick and the Dead: Artists and Anatomy* (Manchester: Cornerhouse Publications, 1997); exhibition curated by Deanna Petherbridge at the Royal College of Art, London; Mead Gallery, Warwick Arts Centre, Coventry; Leeds City Art Gallery, Leeds.
Explores the imagery of anatomical illustrations and focuses on more than 90 artists, including Leonardo da Vinci, Albrecht Dürer, William Hogarth, George Stubbs, Kiki Smith and Cindy Sherman.

Prag, John, and Neave, Richard, *Making Faces: Using Forensic and Archaeological Evidence* (London: British Museum Press, 1997).
John Prag and Richard Neave have performed pioneering work reconstructing the facial appearance of ancient people using the evidence provided by their remains. *Making Faces* explains the historical circumstances surrounding a number of case studies, and describes the search for evidence to recreate a likeness.

Simblet, Sarah, 'The Head', in *Anatomy for the Artist*, (London: Dorling Kindersley, 2001).
Uses drawing and photography to reveal intricate details of anatomical structures, showing the fascinating relationships between internal structures and external appearance.

psychology and neurophysiology

Bruce, Vicki, and Young, Andy, *In the Eye of the Beholder: The Science of Face Perception* (Oxford: Oxford University Press, 1998).
An introduction to the science of the human face and the psychology of face perception.

Cole, Jonathan, *About Face* (Cambridge, Mass.: MIT Press, 1998).
Presents 'a natural history of the face and an unnatural history of those who live without it', drawing on neurology, human development, anthropology, philosophy and the arts, and centred on case studies of facial loss through blindness, autism and neurological impairment.

Ekman, Paul, *Emotions Revealed: Understanding Faces and Feelings* (London: Weidenfeld and Nicolson, 2003).
Describes and explains the universal emotions – sadness, anger, surprise, fear, disgust, contempt and enjoyment.

Partridge, James, *Changing Faces: The Challenge of Facial Disfigurement* (Harmondsworth: Penguin, 1990).

A book intended for sufferers of facial and other disfigurements, and their families, including information about the impact of injuries and accidents on those involved, facial injury and its personal and social consequences, plastic surgery – its risks, benefits and limitations – the routine of hospital life, and the future.

future face

Budd, Robert, Finn, Bernard, and Trischler, Helmuth, eds., *Manifesting Medicine: Bodies and Machines* (Amsterdam: Harwood Academic Publishers, 1999).
Topics covered include technological artefacts as expressions of human culture, the meanings of new technology for medical staff, life and engineering, and comparing living and engineering systems.

Dyens, Ollivier, trans. Evan J. Bibbee and Ollivier Dyens, *Metal and Flesh: the Evolution of Man – Technology Takes Over* (Cambridge, Mass.: MIT Press, 2001).
Explores the transformation of our perceptions by technology, the emergence of a cultural biology and the human/machine entanglement that is changing the way we live.

Ikam, Catherine, and Fléri, Louis, *Portraits Réel/Virtuel* (1999: Paris); virtual installation created by Catherine Ikam for the Ircam-Centre Georges Pompidou, June 1996.
Interview between Catherine Ikam and Paul Virilio on the potential futures of digital technologies and virtual reality.

Menzel, Peter, and D'Aluisio, Faith; *Robo Sapiens: Evolution of a New Species* (Cambridge, Mass: MIT Press, 2001).
Presents the next generation of intelligent robots and their makers through photographs, extensive interviews with robotics pioneers and anecdotal field notes, with behind-the-scenes information and easy-to-understand technical data about the machines.

Wilson, Stephen, *Information Arts: Intersections of Art, Science and Technology* (Cambridge, Mass: MIT Press, 2001).
An investigation of the impact of the digital revolution on art and the provision of new creative and scientific tools.

copyright acknowledgements

picture credits

Cover Michael Najjar, *dana_2.0* (courtesy of Michael Najjar)

1 Michael Najjar, *dana_2.0* (courtesy of Michael Najjar)

8 Franz Xavier Messerschmidt, *The Yawner* (detail) circa 1777–83 (Szépmüvészeti Músеum, Budapest)

10 Mug-shot of Michael Jackson, 2003 (Santa Barbara County Sheriff's Department)

11 Wladyslaw Theodore Benda, Myrna Loy lifemask, circa 1940 (National Portrait Gallery, Smithsonian Institution)

14 André Kertész, *The Puppy*, 1928 (photo André Kertesz © Ministère de la Culture – France)

15 Jo Longhurst, *Terence*, 2003 (courtesy Jo Longhurst)

20 Robert Mapplethorpe, *Ken Moody, 1983* (Ken Moody, 1983 / © The Robert Mapplethorpe Foundation, courtesy Art and Commerce Anthology)

22 Gunter von Hagens, Plastinated vascular system of the head, 2000 (© Gunther von Hagens, Institute for Plastination, Heidelberg, Germany; www.bodyworlds.com)

23 Marc Quinn, *Self*, 1991 (© Mark Quinn. Courtesy of Jay Jopling/White Cube (London))

26 Edwin Romanzo Elmer, *Mourning Picture*, 1890 (courtesy Smith College Museum of Art)

27 Unknown, Photograph of father with mother holding dead child, circa 1850–1860, daguerreotype 1/4 plate (courtesy of Strong Museum)

28 Rudolf Schaefer, from the series *Dead Faces* 1986 (courtesy Rudolf Schaefer)

29 Professor Dr H. Killian, *Sarkom des Schienbeines mit Metestasen in der Lunge Abbildu?? 25* from *Facies Dolorosa* 1967 (Dustri-Verlag Dr. Karl Feistle OHg)

32 Fruitstone amulet (© The ScienceMuseum/Science and Society Picture Library)

32 Brassempuoy Venus (© The Natural History Museum)

33 Plaster skull from Jericho circa 7000–6000 BC (© The Trustees of the British Museum)

36 Camille Silvy, *Adelina Patti as Martha*, 1861 (courtesy National Portrait Gallery)

37 Patricia Piccinini, *Psychotourism*, 1994–5 (courtesy Patricia Piccinini)

41 Charles Bell, 'The muscles of the face', 1824 (Wellcome Library, London)

42 Daniel Lee, *1949 – Year of the Ox* (courtesy Daniel Lee)

43 Catherine Ikam & Louis Fléri, *Alex*, 1995 (courtesy Catherine Ikam & Louis Fléri)

46 James Deville, *William Blake*, 1823 (National Portrait Gallery, London)

48 Antonio Durelli, *Ecorché Head & Neck*, 1837 (Wellcome Library, London)

49 Primal Pictures, still image derived from *Interactive Head & Neck*, part of a computer model of the human anatomy, www.primalpictures.com (courtesy Primal Pictures)

52 Francis Bacon, *Study for a Portrait II (after the life mask of William Blake)*, 1955 (Estate of Francis Bacon 2004. All rights reserved, DACS)

53 J. F. Gautier d'Agoty, *Two Heads with Brains Exposed*, 1748 (Wellcome Library, London)

54 Christian Dorley-Brown, *15 Seconds*, 1994–2004 (© Christian Dorley-Brown)

55 Richard Neave, *Professor Peter Egyedi*, 1996 (courtesy Richard Neave, RN-DS-Partnership, Manchester UK)

55 Richard Neave, *Reconstructed head of the unknown Dutchman*, 1996 (courtesy Richard Neave, RN-DS-Partnership, Manchester UK)

58 Hasbro, My Real Baby, 2000 (courtesy of Peter Menzel/ Science Photo Library)

59 David Hanson, K-Bot, 2003 (courtesy of David Hanson)

60 Guillaume Benjamin Amand Duchenne de Boulogne/ Tournachon; *Mécanisme de la physionomie humaine*, 1862 (Wellcome Library, London)

61 Guillaume Benjamin Amand Duchenne de Boulogne, *Electrisation*, 1862

64 Claude Cahun and Marcel Moore, *MRM (sex)*, circa 1929–30 (Aveux non avenus / Paris: Editions du Carrefour, 1930)

65 Elia Alba, *Doll Heads (Multiplicities)*, 2001 (courtesy Elia Alba/Henrique Faria Fine Art)

68 Henry Tonks, *Studies for Facial Wounds* (reproduced by kind permission of the President and Council of the Royal College of Surgeons of England)

69 Mark Gilbert, *Chris (I) & (ii)*, 1999 (courtesy of Mark Gilbert)

74 Copy of an original photograph of Joseph Clover demonstrating

how chloroform was administrated with his apparatus, English, 1862 (Science Museum)

75 Gunner wearing silver mask (Edinburgh Royal College of Surgeons)

78 & 79 *Repairing War's Ravages: Renovating Facial Injuries. The patient examining the mould of his own face/mask for facial disfigurement* (© The Imperial War Museum)

82 Leprosy demon mask (© The Science Museum/ Science and Society Picture Library)

83 Iron mask (© The Science Museum/ Science and Society Picture Library)

84 Jean Pierre Khazem, *Mona Lisa Live*, 2003 (courtesy of Art + Commerce Anthology)

85 Paddy Hartley, *Face Corset* (courtesy of Paddy Hartley)

88 Helena Rubinstein, Pomade Noir Masque, 1939 (courtesy of *Vogue*, Condé Nast Publications Inc.)

89 Helena Rubinstein, Masque treatment, 1939 (courtesy of *Vogue*, Condé Nast Publications Inc.)

94 Noh mask of a young woman (© The Trustees of the British Museum)

95 Leiter of Vienna, Bakelite face phantom, circa 1860–90 (© The Science Museum/ Science and Society Picture Library)

96 Corrine Day, *Me just before brain surgery, London Hospital, 1996*, 2000 (courtesy of the artist and Gimpel Fils, London)

97 Max Factor Beauty Calibrator, 1932 (courtesy Hollywood Entertainment Museum)

100 Harwood, *White Man Black Mask*, Mongrel 1996 (courtesy of Mongrel)

104 Orlan having cosmetic surgery in New York, 1993 (© Rex Features)

105 *Allegory of Vanity*, circa 18th century (© The Science Museum/ Science and Society Picture Library)

108 G. E. Madeley, after G. Spratt, *Physiognomist*, 1831 (Wellcome Library, London)

109 T. Rowlandson, *Franz Joseph Gall leading a discussion on phrenology*, 1808 (Wellcome Library, London)

112 Cesare Lombroso, Portraits of German criminals, circa 1895 (courtesy Research Library, The Getty Research Institute, Los Angeles)

113 Known militant suffragettes' police ID cards, 1914 (National Portrait Gallery)

115 Francis Galton, Family composite portraits, circa 1883 (Wellcome Library, London)

116 Alphonse Bertillon, Type portraits, circa 1890 (courtesy of George Eastman House, New York)

120 Tibor Kalman, *Black Queen Elizabeth* (courtesy of Colors Magazine; No. 4: Race), 1993

121 Moors murderer, Myra Hindley 1966 (© Rex Features)

122 Christian Dorley-Brown, *Haverhill 2000* (© Christian Dorley-Brown)

123 Christian Dorley-Brown, *Haverhill 2000* (© Christian Dorley-Brown)

128 Biometric mesh (A4Vision Inc.)

129 Heather Barnett, *One Man's Land*, 2002 (courtesy of Heather Barnett)

132 Pierre Huyghe & Philippe Parreno, *Skin of Light*, from the series 'No Ghost just a Shell: Annlee' (Collection Van Abbemuseum, Eindhoven, Netherlands)

133 Daniel Robichaud, *Digital Marlene* (courtesy Daniel Robichaud)

138 Alceu Baptistão, Kaya wireframe (courtesy of Alceu Baptistão)

139 Alceu Baptistão, Kaya (courtesy of Alceu Baptistão)

142 & 143 Man Ray, *Marquise Casati*, 1922 (© Man Ray Trust / ADAGP-DACS/ telimage – 2004)

146 Wanda Wulz, *Cat and I*, double exposure 1932 (courtesy of Archivi Alinari)

148 & 149 Douglas Gordon, *Monster I*, 1996–7 (courtesy of the artist and the Lisson Gallery, London. Collections of Eileen Harris-Norton and Peter Norton, Santa Monica)

152 Johann Kaspar Lavater, *From Frog to Apollo*, circa1855 (Wellcome Library, London)

153 Giambattista della Porta, *Biggest Head*, 1586 (courtesy Research Library, The Getty Research Institute, Los Angeles)

156 Jocelyn Wildenstein (© Rex Features)

157 Boris Karloff as Frankenstein's monster, from *Bride of Frankenstein*, 1935 (© Rex Features)

158 Tattooed face of Paul Morris, 1999 (© Rex Features)

159 LawickMüller (Fredericke van Lawick/Hans Müller), *Athena Velletri – Nina* , from the series

PERFECTLY SuperNATURAL, 1999 (courtesy galerie Patricia Dorfmann, Paris)

162 Lester Gaba, *Cynthia*, New York, 1936 (courtesy of Marsha Bentley Hale)

163 Pierre Imans, *Mannequins*, 1930s (courtesy of Marsha Bentley Hale)

164 David Hopkinson and Shelley Wilson, *Beneath the Mask* (courtesy of David Hopkinson and Shelley Wilson)

165 Cindy Sherman, *Untitled MP 316*, 1995 (courtesy of Cindy Sherman and Metro Pictures Gallery)

173 Alf Linney, Photograph for planning surgery to correct facial anomalies (courtesy of Alf Linney)

174 Alf Linney, Facial scans of a child eight years apart (courtesy of Alf Linney)

175 Alf Linney, Average facial scans (courtesy of Alf Linney)

176 Alf Linney, Family members (courtesy of Alf Linney)

177 Alf Linney, 3D image from CT scans (courtesy of Alf Linney)

177 Alf Linney, Surgical simulation (courtesy of Alf Linney)

178 Alf Linney, Result of simulated surgery (courtesy of Alf Linney)

180 Alf Linney, Details of skull defect using CT scan data (courtesy of Alf Linney)

180 Alf Linney, Prototype plastic model of jawbone (courtesy of Alf Linney)

181 Alf Linney, Mixing images of real and virtual bones (courtesy of Alf Linney)

185 Male and female faces (courtesy of University of Stirling)

186 Adjusting female and male faces (courtesy of Ian Penton-Voak, University of Stirling)

187 Composites of footballers (courtesy of Peter Hancock, University of Stirling)

189 Photofit and E-Fit systems (courtesy of Charlie Frowd, University of Stirling)

190 George Clooney composites (courtesy of University of Stirling)

191 Facial composites (courtesy of University of Stirling)

192 Evo-Fit composites (courtesy of Charlie Frowd, University of Stirling)

196 Elastick skin man (Wellcome Library, London)

197 Anthony Aziz and Sammy Cucher, *Dystopia, Maria*, 1994 (courtesy Aziz + Cucher)

text credits

Permission has been granted for the use of quotations from the following sources:

Kobo Abe, *The Face of Another*, trans. E. Dale Saunders (New York: Vintage, 1966), pp. 31–40.

W. H. Auden, 'Concerning the Unpredictable', *New Yorker*, 21 February 1970, p. 124.

Roland Barthes, *Camera Lucida: Reflections on Photography*, trans. R. Howard (London: Jonathan Cape, 1982), pp. 13, 63–72, 81.

Charles Bell, *Essays on the Anatomy and Philosophy of Expression* (London: John Murray, 2nd edn, 1824), p. xvii.

W. T. Benda, *Masks* (New York: Watson Gupthill Publications Inc., 1944) pp. vii–viii, 24.

Marsha Bentley Hale, 'Lasting Expressions', *Visual Merchandising and Store Design*, November 1985, p. 45.

Alphonse Bertillon, quoted in Christian Pheline, 'L'image accusatrice', in *Les Cahiers le la Photographie No 17* (1985), p. 129.

Nathanial Burgess, "Taking Portraits after Death', from *The Photographic and Fine-Art Journal* Vol. 8, No. 3 (March 1855) pg. 80.

Angela Carter, 'The Tiger's Bride', in *The Bloody Chamber and Other Stories* (first published by Gollancz, 1979, published by Vintage, 1995). © Angela Carter 1979.

Emma Chambers, *Henry Tonks: Art and Surgery* (London: UCL, 2002), p. 16.

Jonathan Cole, *About Face* (Cambridge, Mass: MIT Press, 1988), pp. 2, 43–52, 96, 192.

Anna Coleman Ladd, 'How Wounded Soldiers Have Faced the World Again with "Portrait Masks" ', interview by P. Kind, *St Louis Post-Dispatch*, 26 March 1933. Reprinted with permission of the *St Louis Post-Dispatch*, © 1933.

Guillaume Benjamin Amand Duchenne de Boulogne, *The Mechanism of Human Facial Expression*, ed. and trans. R. Andrew Cuthbertson (Cambridge: Cambridge University Press, 1990), p. 19.

Jenny Diski, *The Dream Mistress* (Guernsey, Channel Islands: Phoenix, 1996), pp. 16-17. Reprinted with permission of A. P. Watt Ltd on behalf of Jenny Diski.

Bob Dylan, 'Sara', words and music by Bob Dylan © Ram's Horn Music, Sony/ATV Music Publishing Ltd.

All rights reserved. International copyright secured.

Paul Ekman, *Emotions Revealed: Understanding Faces and Feelings* (London: Weidenfeld and Nicolson, 2003), p. 15.

Ricardo Galeazzi, 'Rebuilding disabled soldiers: wonderful work that Italy is doing to render maimed men self-supporting', in *Current History* (New York: July 1918), .p. 101.

Charles Mills Gayley, *The Classic Myths: In English Literature and in Art* (Boston: The Athenaeum Press, 1911), p. 208.

Stephen Jay Gould, *The Mismeasure of Man* (New York: W. W. Norton and Company, 1981), p. 174.

Philippe Halsman: A Retrospective (Bullfinch Press, Little Brown, 1998).

Peter Hamilton and Roger Hargreaves, *The Beautiful and the Damned: Nineteenth-Century Portrait Photography* (London: The National Portrait Gallery in association with Lund Humphries, 2001), pp. 49, 96.

David Hanson, personal correspondence, 2004.

Mark B. N. Hanson, 'Affect as Medium, or the "digital-facial-image" ', *Journal of Visual Culture*, 2,2 (August 2003), p. 214.

Nathaniel Hawthorne, 'The Prophetic Pictures', *Twice Told Tales and Other Short Stories* (London: Readable, 1842), p. 23; as quoted in Heinz K. and Bridget A. Hemisch, *The Photographic Experience 1839–1914* (Pennsylvania: Penn State Press, 1994), p. 166.

Malvina Hoffman, *Sculpture Inside Out* (London: G. Allen and Unwin, 1939), p. 135.

Catherine Ikam and Louis Fléri, *Portraits Réel/Virtuel* (catalogue; 1999: Paris); pp. 27–8.

Claudia Johnson, 'Fair maid of Kent? The arguments for (and against) the Rice portrait of Jane Austen', *Times Literary Supplement*, 13 March 1998, p. 15.

From *Paul Klee*, exhibition catalogue (London, Marlborough Fine Arts, 1966), no. 45; quoted in Richard Brilliant, *Portraiture* (London: Reaktion, 1991), p. 171.

Susan Kuchinska, 'Image is Everything', *Wired*, 18 June 1998.

Milan Kundera, *Immortality*, trans. Peter Kussi (London: Faber and Faber, 1991), pp. 34, 35–6.

Johann Kaspar Lavater, *Essays on Physiognomy: Designed to promote the Knowledge and the Love of Mankind*, trans. Thomas Holcroft, 9th edn (London: William Tegg, 1855), pp. 10–11.

Gaston Leroux, *The Phantom of the Opera*, introduction by Max Byrd (New York: New American Library, 1987), p. 9.

John Liggett, *The Human Face* (London: Constable, 1974), p. 275.

Al Lingis, 'The tact and tenderness of the light', unpublished paper delivered at the National Portrait Gallery 'Faciality' conference, December 2003.

Cesare Lombroso, quoted in I. Taylor, P. Walton and J. Yound, *The New Criminology: For a Social Theory of Deviance* (London: Routledge & Kegan Paul, 1973) p. 325.

Paolo G. Morselli, 'The Minotaur Syndrome: plastic surgery of the facial skeleton', *Aesthetic Plastic Surgery*, 17: 99–102, 1993.

Cpl. Ward Muir, *The Happy Hospital* (London: Simpkin, Marshall, Hamilton, Kent & Co. Ltd, 1918), p. 143.

Cpl. Ward Muir, 'The Men with New Faces', *The Nineteenth Century*, October 1917, p. 752.

Erin O'Connor, *The Love Song of Plastic Surgery*, erinoconnor.org/writing/plastics.shtml

George Orwell, *Nineteen Eighty-four*, 1949 (Harmondsworth: Penguin, 1983), p. 230. Reprinted with permission of Bill Hamilton as the Literary Executor of the Estate of the late Sonia Brownell Orwell and Secker & Warburg Ltd.

Philippe Parreno and Pierre Huyghe, quoted in Arne Altena, 'Between illusion and artificiality: computer animated heads', www.outlineamsterdam.nl/nlprogramma/2003

Patricia Piccinini, *The Mutant Genome Project*, 1994-95, from www.patriciapiccinini.net

Francis Pierce, from Alberto Manguel, *Bride of Frankenstein* (London: BFI, 1997), pp. 20–21.

Marcia Pointon, 'Kahnweiler's Picasso; Picasso's Kahnweiler' in Joanna Woodall (ed.), *Portraiture: Facing the Subject* (Manchester and New York: Manchester University Press, 1977), p. 190.

Reginald Pound, *Gillies: Surgeon Extraordinary. A biography* (London: Michael Joseph, 1964).

Rainer Maria Rilke, *The Notebooks of Malte Laurids Brigge*, introduction by Stephen Spender, translation John Linton (Oxford: Oxford University Press, 1984), p. 6. Rights granted by Random House, London.

Quoted in Alan Clive Roberts, *Facial Prosthesis: The Restoration of Facial Defects by Prosthetic Means* (London: Henry Kimpton, 1971), p. 1.

Jay Ruby, *Secure the Shadow: Death and Photography in America* (Cambridge, Mass: MIT Press, 1995), p. 179.

Allan Sekula, 'The Body and the Archive', *October*, 39 (Winter 1986), pp. 3–64.

Sarah Simblet, 'The Head', in *Anatomy for the Artist* (London: Dorling Kindersley, 2001), pp. 47–57.

Daniel Smith, 'Gilles Deleuze: essays critical and clinical', trans. Daniel W. Smith and Michael A. Greco (Verso, 1998), p. xxxii.

David Sylvester, *The Brutality of Fact: Interviews with Francis Bacon*, 3rd enlarged edn (London: Thames and Hudson, 1987), p. 50.

A. W. Young, D. C. Hay and A. W. Ellis, 'The faces that launched a thousand slips: everyday errors and difficulties in recognising people', *British Journal of Psychology*, 76, 495–523.

Every effort has been made to obtain copyright permission for quoted extracts. Any omissions or errors of attribution are unintentional and will, if brought to the attention of the publisher, be corrected in future printings.

To claim your copy of the CD-ROM catalogue and booklet which accompany the *Future Face* exhibition, please write to:

Marketing Department
The Wellcome Trust
FREEPOST
ANG 6754
Ely CB7 4BE

or email: marketing@wellcome.ac.uk

The CD-ROM and booklet will be sent to you free of charge, subject to availability. The Marketing Department at the Wellcome Trust may also be contacted via telephone or fax:
Tel: +44 (0)20 7611 8651
Fax: +44 (0)20 7611 8416

The Wellcome Trust is an independent research-funding charity, established under the will of Sir Henry Wellcome in 1936. Its mission is to foster and promote research with the aim of improving human and animal health. Its work covers four areas:

Knowledge – improving our understanding of human and animal biology in health and disease, and of the past and present role of medicine in society;

Resources – providing exceptional researchers with the infrastructure and career support they need to fulfil their potential;

Translation – ensuring maximum health benefits are gained from biomedical research;

Public engagement – raising awareness of the medical, ethical, social and cultural implications of biomedical science.

wellcometrust